Great Buildings of the World
Bridges
Derrick Beckett

Paul Hamlyn
London.New York.Sydney.Toronto

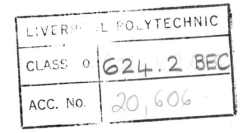
Published by The Hamlyn Publishing Group Limited
London . New York . Sydney . Toronto
Hamlyn House, The Centre, Feltham, Middlesex

© The Hamlyn Publishing Group Limited 1969
SBN 600 01640 4

Printed and bound in Great Britain by
Thomas Nelson (Printers) Ltd., London and Edinburgh

Contents

Introduction

The function of a bridge is to provide a passageway for people, vehicles or materials, both solid and liquid, where normal surface construction is not practical. It is this singleness of purpose that distinguishes a bridge from other forms of structure. The design of building structures, for example, involves several variables: spatial requirements, heating, ventilation, lighting and so on, and in many instances, structure will not be the dominant factor. In a bridge, structure dominates and its form should reflect this. The essence of an elegant bridge is simplicity of line, in which the structural form is expressed to the full. The addition of superfluous features will in general detract from the beauty of a bridge rather than enhance it. This simplicity of line is as necessary in the design of a simple foot-bridge as in the structure of a bridge for crossing the English Channel.

Brunel's design for the Clifton Bridge at Bristol (above left) illustrates how simplicity of line immeasurably enriches the appearance of a bridge. In comparison, Telford's design with its colossal Gothic-style piers clashes disastrously with the superb landscape of the Avon gorge. A bridge should be designed to blend with the landscape rather than compete with it. It has been said that it is difficult to make a bridge look ugly, but this has often been achieved by the incorporation of heavy parapets and unnecessary ornament.

For an engineer the prime object in the design of any structure is that it should transmit loads to the ground with an adequate margin of safety against collapse or excessive deformation. The significant loads to be considered in the design of a bridge are the self-weight of the structure, which is invariable, and the vehicular and pedestrian loads, which are variable. In certain types of bridge structure, the effect of wind loading must also be investigated.

In many bridges the self-weight of the structure is the principal loading and in large spans may exceed 80 per cent of the total load to be transmitted to the ground. Technological developments in the last 100 years have enabled engineers to reduce the material content, and so the self-weight of bridges, even though vehicle loads have increased. The continual increase of heavy loads such as generators and transformers travelling by road has meant that bridges on major roads have to be designed to accept single vehicle loads of over 200 tons.

Brunel's design for the Clifton Bridge at Bristol (above left) blends well with the beautiful Avon gorge.

In contrast to Brunel's design, Telford's proposal (below) with its Gothic-style piers clashes disastrously with the landscape.

8

Simplicity of line is an essential feature of good bridge design – in a simple footbridge (left) no less than a planned bridge across the English Channel with a span of several thousand feet.

The design standards for bridge works of any importance in Great Britain are laid down by the Ministry of Transport. These standards are aimed at ensuring that new bridges will have a useful life of 100 to 200 years—in other words, guaranteeing that the structure is durable and has an adequate margin of safety against failure.

Say a bridge designed to carry a total load of 500 tons collapsed at a 10 per cent overload, the margin of safety against failure would be completely inadequate. Most bridges are designed to carry a considerable overload (vehicular loading) and would become unserviceable to traffic well before the collapse load is reached. At a load of, say, 70 per cent of the collapse load the deformation of the structure would probably be so great that the surface of the road would be seriously impaired. One of the Ministry of Transport design standards stipulates that a bridge must have a minimum collapse load of 1·5 times the self-weight plus 2·5 times the pedestrian and vehicular loading. For large spans this ensures a high margin of safety, since the self-weight may be several times the vehicular loading. Design standards vary throughout the world, so it is difficult to compare the safety margins of bridge structures in 9

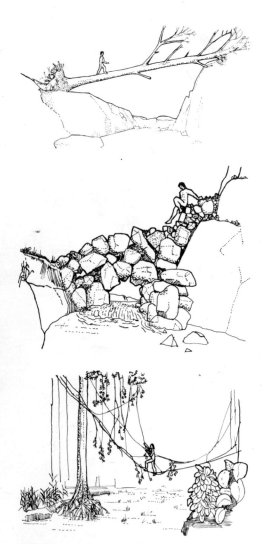

All bridges are based on three structural forms developed from Nature: the beam bridge from a fallen tree; the arch bridge from a rock fall; the suspension bridge from a hanging vine or creeper.

different countries. But every bridge, wherever it is built, is based on one or more of three basic structural forms, which were all developed from nature: the beam bridge from a fallen tree; the arch bridge from natural rock formation; and the suspension bridge from a hanging vine or creeper. These three forms have remained unchanged for thousands of years, and it is only recently that man's knowledge of structural analysis and the response of materials to force actions has enabled their load-carrying capacity to be predicted with any degree of accuracy. Before the 16th and 17th centuries theories of structural behaviour based on combined mathematical and experimental investigations were practically unknown.

The loads acting on a structure produce five basic types of force action—compression, tension, bending, shear and torsion. These are illustrated in figure 1. Force is normally associated with muscular effort but in structural language, force is defined in magnitude (pounds, tons, kilograms and so on) and direction. A vehicle weighing 10 tons moving over a bridge (figure 2) can be represented by a force of magnitude 10 tons whose direction is vertically downwards. The weight of the bridge deck itself may be represented by a series of forces (figure 3) acting vertically downwards. The braking of a vehicle moving across a bridge could be represented by a horizontal force acting at the roadway level (figure 4). It should be noted that for a bridge deck of constant depth its self-weight can be represented by a loading that is uniformly distributed along its length (figure 5); this would also apply approximately to the vehicular loading shown in figure 6. The shape of many bridge structures remains constant whether they are loaded or not—that is, deformations are negligible. (Structures which do not remain constant under load will be considered later.) The structure may thus be considered as a rigid body and must be in equilibrium under the forces acting on it. The effect of the forces will be to produce movement, rotational and translational. Consider the equilibrium of a beam bridge—for example, the Post Bridge, which spans the River Dart in the centre of Dartmoor. The weight of the granite slabs (figure 7) can be represented by arrows, which indicate the direction of the forces acting. To produce equilibrium—that is, to prevent the slabs falling into the river—forces in the opposite direction must be applied (reactions), which are provided by the supporting granite piers. The loads acting on the piers are in turn transmitted to the ground. If one of the piers was removed, the structure would

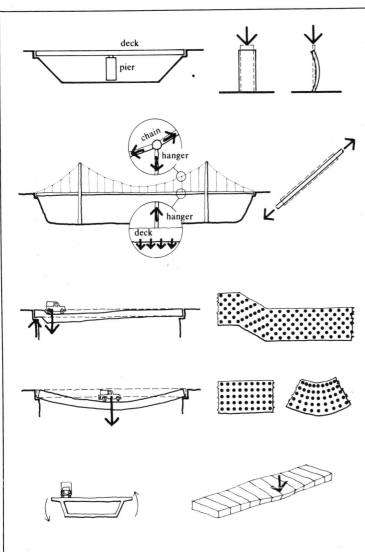

Fig. 1. The five basic types of force action are (top to bottom): compression, causing shortening or buckling of a member such as a bridge pier; tension, causing extension of a member such as a suspension-bridge chain or hanger; shear or racking of a member such as a bridge beam; bending, extension and compression of a member such as a bridge beam; torsion or twisting of a bridge deck due to heavy loading at its edge.

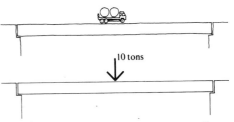

Fig. 2. A vehicle load can be diagramatically represented by an arrow indicating its direction and a number showing its magnitude.

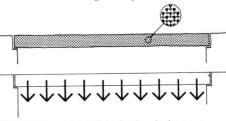

Fig. 3. The weight of the bridge deck structure can be represented by a series of arrows pointing vertically downwards.

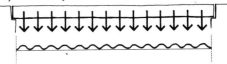

Fig. 4. The braking force of a vehicle can be represented by a horizontal arrow.

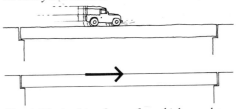

Fig. 5. For a deck of constant depth the self-weight can be represented by a uniform loading, indicated by arrows or a wavy line.

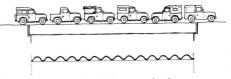

Fig. 6. A string of vehicles can also be represented approximately by a uniformly distributed load.

11

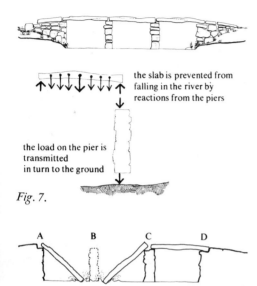

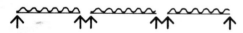

the slab is prevented from falling in the river by reactions from the piers

the load on the pier is transmitted in turn to the ground

Fig. 7.

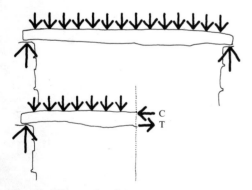

A   B   C   D

Fig. 8. If pier B is removed the structure will collapse by rotation about points A and C.

Fig. 9 A mathematical model of Post Bridge, Devon. The weight of the granite slabs is represented by a uniformly distributed load.

C
T

Fig. 10. The weight of the slabs produces a bending effect, causing compression (C) in the top slab and extension (T) in the bottom slab.

12

collapse by rotation about points A and C (figure 8). A mathematical model of this structure, which represents the simplest form of beam bridge, is shown in figure 9. It can be seen that the force actions set up by the weight of the slabs, which are about 15 feet long, will produce a bending effect causing compression in the top of the slab and extension in the bottom (figure 10). At this point, it is useful to introduce the term bending moment (figure 11). The bending moment at any point on a structural member induced by a force acting on it is the product of the force multiplied by the perpendicular distance of its line of action from the point considered. The larger the bending moment the greater the amount of compression and extension of the fibres of a beam. Thus if the distribution of bending moment can be plotted along the length of a beam the positions of maximum extension and compression of the fibres in the beam can be determined (figure 12).

The resistance of materials to compression and extension varies. Stone, concrete and cast iron have little resistance to extension whereas modern steels have a high resistance to both compression and extension. At the time the Post Bridge was built man had no quantitative knowledge of structural theory and so relied on intuition and previous experience. The placing of these large granite slabs in position represented a formidable construction problem. However, in China during the early middle ages a number of stone beam bridges were built on a scale that dwarfs the Post Bridge. These consisted of granite beams of up to 70 feet in length, weighing around 200 tons, and it is difficult to imagine how they were ever placed in position. Other structural forms—the arch and chain—also have a long history, but before considering them, the development of beam bridges will be outlined.

The principles of equilibrium were first outlined by Archimedes (287–212 BC) and no doubt some simple rules of thumb were developed early on for proportioning structural members. The Romans did not in general use beams as structural members for bridges and the first tests to determine the load-carrying capacity of beams were probably carried out by Leonardo da Vinci (1452–1519), although he made no definite statement about the effect of depth on the strength of beams.

The first mathematical expressions for the strength of beams in bending were developed by Galileo (1564–1642). These investigations into the strength of materials were carried out when he was teaching at the University of Padua. (The structures built in Galileo's time, whether bridges or buildings, generally had their own weight as the dominant load and often collapsed because they were unable to support themselves.) The principal findings of Galileo's investigations were as follows: (a) The strength of a member that is extended longitudinally (extension of a member is produced by a tensile force) is proportional to the surface area of the member and independent of its length (figure 13). This he referred to as the absolute strength. (b) The strength of a member in bending is proportional to the square of the depth for a given width. Galileo assumed that the whole of the section was in tension because of bending (figure 14), which is at variance with that shown in figure 10. (c) The size of structures cannot be increased to vast dimensions because they would break down under their own weight. In other words, the cross-section of a beam must be increased in a greater proportion than that of the spans so that the beams shall keep the same relative strength. The smaller the body the greater its relative strength. Thus a small dog could probably carry on its back two or three dogs of its own size, but a horse could not carry another horse—the weakness of giants. (d) Hollow beams are employed in art and nature for the purpose of increasing strength without increasing weight.

Galileo was the founder of modern scientific research and in some ways his theories seem more up-to-date today than they did in the last century. It is only recently that the concept of absolute strength—statement (a)—has been used as a basis for proportioning structural members.

Experiments by Robert Hooke (1635–1703) and Mariotte (1620–84) led to further expressions being developed for the strength of beams in flexure, which are compared with Galileo's results in figure 15. Hooke, in the manner of his age, published his invention in the form of an anagram *ceiiinossssttuv*, which when solved reads *ut tensio sic vis*, the Latin for 'as the extension so is the force'. The linear relationship between the load and extension (or stress and strain, figure 17) to a large extent forms the basis of modern structural analysis, although this law breaks down at high-stress levels, and indeed for some materials does not apply at all (figure 18).

By altering the support system for a beam the bending moment distribution can be varied. Various arrangements of beam systems used for bridge construction are shown in figure 16. By making a beam continuous over a number of supports, the maximum values of the bending moment can be reduced for a given load and span.

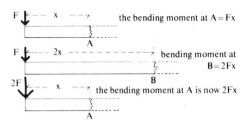

Fig. 11. A bending moment can be expressed mathematically by the product of force (F) times distance (x).

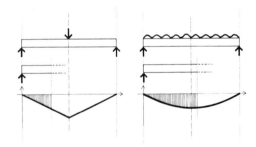

Fig. 12. The variation in bending moment along the length of a beam can be plotted, and thus the position of maximum extension and compression can be determined.

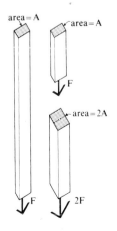

Fig. 13. One of the findings of Galileo's investigations into the strength of materials was that the strength of a member extended longitudinally is proportional to its sectional area (A) and is independent of its length.

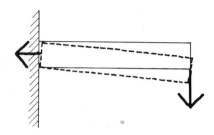

Fig. 14. Galileo assumed that for a member in bending the whole of the cross-section of the beam was in tension at any point.

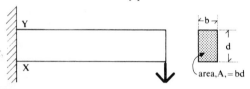

the distribution of stress in section XY according to:

Galileo

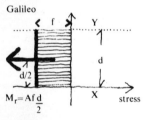

$$M_r = Af\frac{d}{2}$$

Mariotte

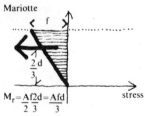

$$M_r = \frac{Af}{2}\frac{2d}{3} = \frac{Afd}{3}$$

Hooke

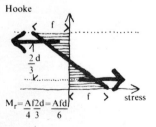

$$M_r = \frac{Af}{4}\frac{2d}{3} = \frac{Afd}{6}$$

M_r = the resistance moment in the beam, which counteracts the applied bending moment

f = the ultimate stress of the material of the beam in tension

Fig. 15.

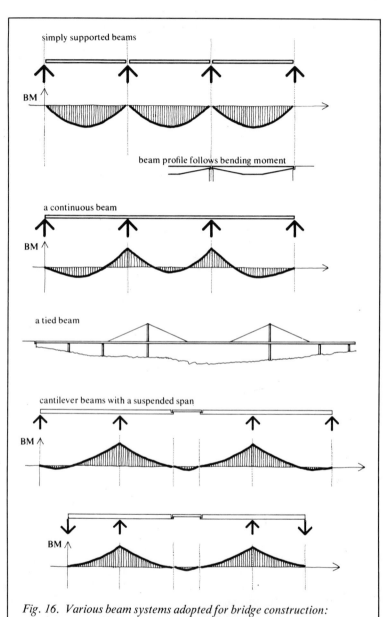

simply supported beams

beam profile follows bending moment

a continuous beam

a tied beam

cantilever beams with a suspended span

Fig. 16. Various beam systems adopted for bridge construction: simply supported beams; continuous beam and tied beam, which have similar bending moments; and cantilever beams with suspended spans. The merits of these systems are discussed in subsequent chapters.

In long spans where self-weight dominates, the beam profile can be adjusted to take the form of the bending-moment diagram and the section can be made hollow. The beam flanges (figure 19) resist the bending effect and the webs, the shearing effect. Hollow sections are also capable of resisting twisting effects caused by heavy vehicle loads acting near the edge of the bridge deck.

To reduce weight, generally at the expense of increased structural depth, open-web or trussed beams are frequently used for bridge construction. The principle of the trussed beam (figure 20) is that the top and bottom booms resist bending by compressive and tensile forces acting at some distance apart while the vertical and inclined members resist the shear. Thus a relatively light section is produced, and it is normally assumed that all the members are subject to direct compression or tension. The truss is in effect an assembly of bars that can be made up from relatively small sections. When used for long spans, the overall depth of the truss becomes very large and the compression members may in fact have to be hollow sections to guard against buckling. Timber has been used for trusses since Roman times although in the past 100 years it has gradually been replaced by iron and steel. However, in countries where there is a plentiful supply of good-quality timber, the development of timber technology, especially in the design of the connections between the members, enables timber to compete even now with steel and concrete for small- and medium-span bridges.

To turn now to the suspension bridge, in which a continuous cable supports the deck by means of suspenders (figure 21). Since the distance between the suspenders is small compared with the overall length of the structure, the self-weight of the deck can be kept to a minimum. Also, because the cable is in tension and materials such as rope, bamboo and steel have a much higher resistance in tension for a given cross-sectional area and length than they have in compression, this form of structure is the most economical for bridge spans of more than 2000 feet. The suspension bridge is of great antiquity and was developed many centuries ago in China, India, Africa and South America; the materials first used for the cables were plants of the liana variety and bamboo. Large-span suspension bridges using iron and steel for the cables evolved in the 19th century in North America and Europe.

The basic difference between the suspension bridge and other bridge systems is that it is likely to change shape under variable loads

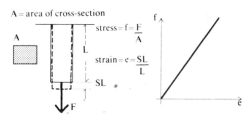

Fig. 17. Hooke's Law, which still forms the basis of modern structural analysis, states 'as is extension so is the force.' The graph illustrates the linear relationship between load (stress) and extension (strain).

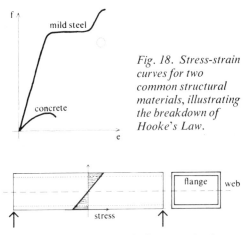

Fig. 18. Stress-strain curves for two common structural materials, illustrating the breakdown of Hooke's Law.

Fig. 19. For large spans the beam section is made hollow to reduce self-weight. The beam flanges resist the bending effect and the webs resist shear.

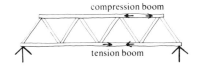

Fig. 20. A solid-web beam can be replaced by a deeper open-web or trussed beam.

15

Fig. 21. A suspension bridge. The deck is supported by a continuous cable, the load being transmitted to the cable by means of suspenders (hangers).

Fig. 22. The angle of a cable will alter for different positions of the load.

Fig. 23. If the loading is uniformly distributed and the weight of the cable is small compared with that of the deck, the cable will take a parabolic form.

(figure 22). The passage of a moving load across the bridge causes continuous alteration to the shape of the chain resulting in considerable deformations and oscillations in the roadway. Only when the moving load is small compared with the weight of the roadway could a simple suspension bridge be used with safety. If the weight of the chain is slight compared with that of the roadway and the roadway load is uniformly distributed along the length of the bridge, the curve assumed by the chain will be parabolic (figure 23). To accept moving loads a suspension bridge must be stiffened to ensure that it maintains its shape. This is normally achieved by stiffening the roadway by means of a trussed or box girder (figure 24).

The structure will also change its shape because of wind loading. This is the principal cause of failure; a classic example was the failure of the Tacoma Narrows Bridge, Washington, in 1940. This suspension structure with a span of 2800 feet collapsed four months after it was completed. The width between the cables was 39 feet and the roadway was stiffened by 8-foot-deep plate girders. The ratio of girder depth to span was 1/350 compared with the then commonly accepted value of 1/50 to 1/100. When a steady wind blows against an obstacle (for example, a solid-web stiffening girder), its wake takes the form of a so-called vortex street (figure 28). The shedding of these vortices on the leeward side of the bridge causes forces to act at right-angles to the direction of the wind, first from one side and then from the other. At a steady wind speed of 42 mph the deck of the Tacoma Bridge twisted about 45 degrees from the horizontal in both directions until it broke. The collapse was filmed by an eye-witness and is a horrifying sight. The disaster led to considerable research into suspension-bridge failure and to the evolution of proper aerodynamic design. A product of this research is the Severn Bridge, which has a span of 3240 feet between piers, a hollow steel box deck 10 feet deep and 75 feet wide and the suspenders inclined to increase stability (figure 25).

The arch is the most familiar of all structural forms used for bridge construction; this type of construction possibly evolved from observations of natural rock falls. The use of stone for beam bridges has obvious limitations since its resistance to flexure is poor. Also, the problems involved in the cutting and handling of large stone slabs are formidable. The arch is essentially a compression structure and is the inverse of the suspension chain. Thus if the suspension chain were inverted (figure 26) it would be subjected to compression,

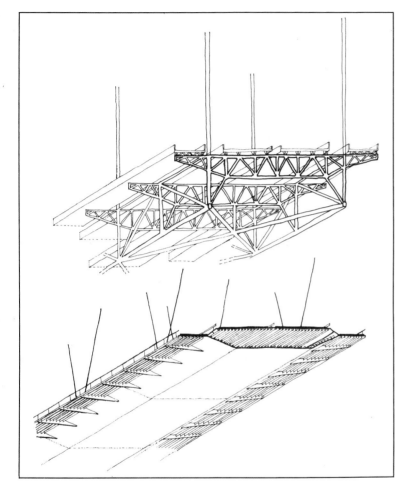

Fig. 24. Trussed girders have been most commonly used to stiffen the deck. A more recent development is the box section (below).

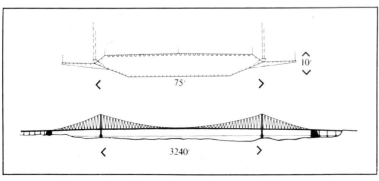

Fig. 25. An elevational sketch of the 3240-foot Severn Bridge and a section through the 10-foot-deep hollow steel-box deck. The suspenders were inclined to increase stability.

17

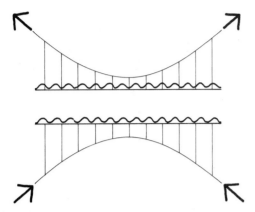

Fig. 26. *An arch can be seen as the inversion of a chain and is essentially a compression structure.*

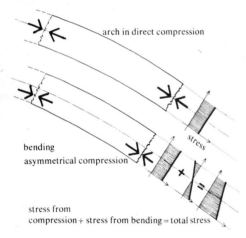

arch in direct compression

bending
asymmetrical compression

stress

stress from
compression + stress from bending = total stress

Fig. 27. *In general some bending will be induced in an arch structure; its effect is illustrated in this sketch.*

Above right, the Tacoma Narrows Bridge as it collapsed, falling 200 feet into the water. The car which can just be seen on the right fell into the river along with the bridge. Note the inclination of the lamp posts.

Fig. 28. *Diagrammatic representation of a vortex street, which causes the deck to oscillate.*

18

not tension, and would collapse. However, if a chain is replaced by a material of such proportions that the compression can be resisted without buckling (which would occur with a slender chain), then a satisfactory structural system will result. It should be noted that the arch will be subjected to a direct compression for the uniformly distributed loading shown if it is parabolic in form. Any other loading will result in bending, which is analogous to the changes of shape of a cable for differing loading conditions. Thus the ideal form for an arch subjected to uniformly distributed loading over the whole of its span is a parabola. Since the loading will not generally be of this form the direct compression will be supplemented by some bending (figure 27). This means that the line of action of the compression force acting on the section (thrust) will be shifted. The profile of an arch is selected to ensure that this shift is kept to a minimum and tension will therefore not develop.

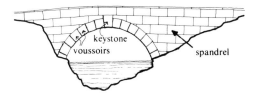

Fig. 29. The principal components of a Roman arch.

From Etruscan and Eastern precedents the Romans developed considerable skill in constructing arches and many of their structures remain intact today. However, they were certainly not able to calculate the strength of their arches although their knowledge of geometry inherited from the Greeks led them to adopt semicircular spans. Their arches were constructed of stone or brick made into the form of wedge-shaped segments or voussoirs (figure 29), built up side by side on temporary timber supports until the top piece at the crown of the arch is pushed home (keystone). A wall (spandrel) was then built up to roadway level. This method of construction is still used for arches although today stone has been replaced by concrete. In the 1000-foot span Gladesville Bridge in Sydney, completed in 1963, the arch ribs were made up from precast concrete voussoirs of a hollow-box section. In this structure (right) the shift of the line of action of the total compressive force in the rib is only $\frac{3}{8}$ inch.

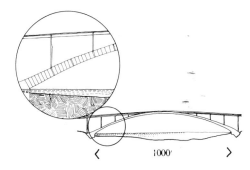

The 1000-foot-span Gladesville arch in Sydney, Australia. The arch ribs were formed of concrete voussoirs of hollow box section to reduce self-weight.

It was not until the 18th century that the theory of arches was developed by French engineers and the idea formulated—which is still widely held—that the tension in a voussoir arch (figure 30) is disastrous. This is not true, however, and even if a joint between two voussoirs opens, failure will not necessarily result. The failure of an arch rib due to the formation of hinges is shown in figure 31.

The beam, arch and chain represent the basic structural forms used for bridges and in all three instances the loads must be transmitted to the ground so that the structure is not overstressed. With the exception of rock—which is the ideal material on which to found a bridge—the resistance of the ground to deformation is very much

19

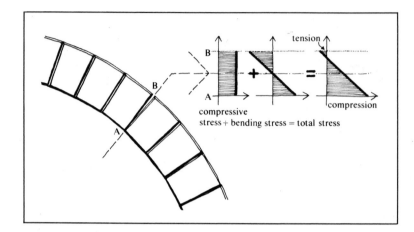

Fig. 30. *The opening of a joint between voussoirs, due to tension developing in one face.*

compressive
stress + bending stress = total stress

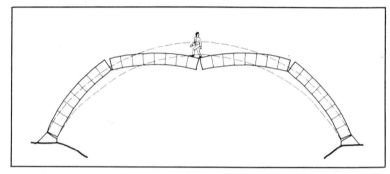

Fig. 31. *The failure of an arch rib through the formation of hinges.*

less than the material used for the bridge deck and piers. The bridge foundations must therefore be of such a form that the loads will be spread over a large area, so reducing the stress on the ground. The foundations must be considered as part of the structure and the nature of the ground will influence the type of structural system selected. To construct an arch structure economically a firm stratum, preferably rock, is needed, since the ground will be subjected to forces that tend to spread the feet of the arch, as well as vertical forces. Foundation systems for various different situations are illustrated in figure 32.

Since the ground conditions can differ considerably from one foundation point to another, the resistance of deformation will also vary. This means that one pier may settle more than another, so amplifying the stresses in the structure. A means of overcoming this is to use a series of simple spans that are not connected.

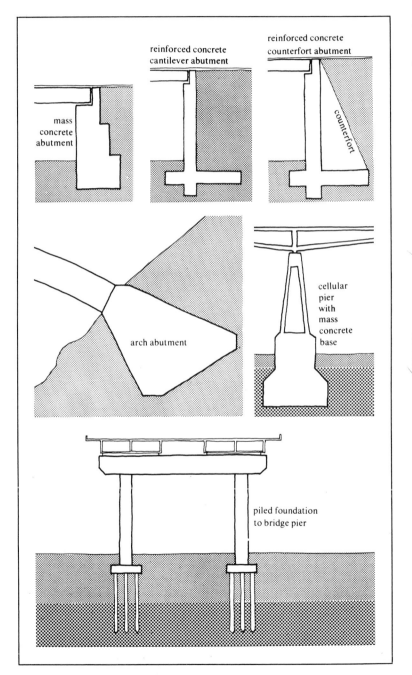

mass
concrete
abutment

reinforced concrete
cantilever abutment

reinforced concrete
counterfort abutment

counterfort

arch abutment

cellular
pier
with
mass
concrete
base

piled foundation
to bridge pier

Fig. 32. Methods of transmitting the load from the bridge deck to the ground: in each case the load is spread over sufficient area to avoid overstressing the soil.

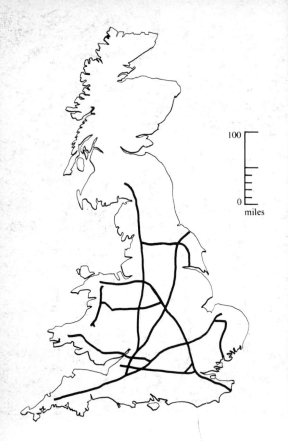

Map showing probable main trackways in early Britain, indicating the need for large numbers of simple bridges of the clapper type.

Above right, stepping stones, Ambleside. The clapper bridge was a development from stepping stones.

Thus settlement of one pier can take place without influencing the stresses in the bridge deck as a whole. Inadequate foundations are the most common reason for structural failures and the construction of modern bridges is preceded by a thorough site investigation, including the testing of numerous soil samples. From this investigation the most economical form for the structure can be worked out.

Water has always been one of the principal obstacles to the improvement of communication by land. In primitive times travellers merely waded through shallow streams or leapt over a series of rock outcrops (stepping stones). In some cases a fallen tree would conveniently span a stream or a rock fall would provide a natural arch. In warmer climates intertwined vines formed natural suspension chains on which men could swing across a river or gorge. Thus the beam, arch and chain bridge all probably developed from natural bridges. Each of these man-made forms has been traced back several thousand years, though which type developed first is a matter of speculation. Arch bridges have been recorded as far back as 4000 BC, and it is likely that crude forms of beam bridge were developed even

Post Bridge, Devon, an early medieval clapper bridge of a type dating from much earlier times.

earlier in which logs or slabs of stone were used to span outcrops of rock. A historic example of a form of beam bridge is that constructed on floating supports (pontoons) by Xerxes, the King of Persia, across the Hellespont at Abydos in 481 BC. Clapper bridges in England—a development from stepping stones—date from early times, though the Post Bridge, Devon, has been proved to be relatively recent. Examples of this form of construction are to be found in other parts of the world, in particular China.

The first organised bridge building on a large scale took place in association with military campaigns. During the expansion of the Empire the Romans built a large number of bridges in stone and timber to carry roads and water (aqueducts). Although the Romans had no means of calculating their structures they were excellent constructors and were able to direct large labour forces supervised by technical experts. They developed an extensive network of roads in Europe; their bridges were about 15 feet wide to allow two vehicles or two legions to pass each other. Almost nothing remains of Roman bridge building in Great Britain except for a few decaying stone arches but there are several fine examples on the continent.

The famous Pont-du-Gard aqueduct near Nîmes in southern France. The structure had an overall length of about 900 feet and consists of three tiers of semicircular stone arches. The projecting stones on the arch faces were used to support wooden staging upon which the arch rings were built.

Perhaps the most famous is the Pont-du-Gard aqueduct near Nîmes in the south of France. This structure consists of three tiers of semi-circular arches. The top tier is 155 feet above the river and has an overall length of about 900 feet. Another example is the bridge across the Tagus at Alcantara in Spain. The length of this bridge is 650 feet with a level roadway throughout. It consists of six arches—the two in the centre of about 100 feet span, those on either side of 60 feet and the two outer of 50 feet. In cities the Romans felt that their bridges should be enhanced by ornament. This use of decoration is exemplified by the Ponte di Augusto at Rimini in Italy, which has five arches, the three in the centre of an equal span of 28 feet and the two outer of 23 feet.

Timber was also widely used by the Romans for bridge construction. The illustration on page 26 shows the wooden superstructure of the bridge over the River Danube, which was erected by order of the Emperor Trajan, as it appears in bas-relief on the Trajan column in Rome. It is claimed that Caesar's engineers built a wooden trestle bridge over the Rhine in ten days—an early example of industrialised building.

From an engineering viewpoint the most interesting aspect of Roman bridge building is the construction techniques that they adopted. The large stones for the arch rings were moved by wedges,

The Roman Bridge across the Tagus, Alcantara, Spain, which consists of six semicircular arches with an overall length of 650 feet. The two central arches have a span of about 100 feet.

Below, the bridge built by Augustus and Tiberius at Rimini, Italy, which still supports heavy traffic. Note the extensive use of ornamentation in contrast to the preceding illustrations.

ramps or hoisting gear. The voussoirs were connected by mortar or iron clamps or sometimes dressed to fit. The voussoirs for the first two tiers of the Pont-du-Gard were dressed to fit and mortar was used to connect the top-tier voussoirs. The projecting stones on the face of the Pont-du-Gard arches were used to support wooden falsework upon which the arch ring could be built. Vitruvius in his books on architecture describes how the Romans constructed the foundations to their buildings and bridges. 'For foundations in water, stones were thrown in and built up to the surface or alternatively

The wooden superstructure of the bridge over the River Danube as it appears in bas-relief on Trajan's Column in Rome.

timber piles were driven to form a cofferdam within which the foundation was built using stones or concrete . . . Then, in the place previously determined, a cofferdam, with its sides formed of oaken stakes with ties between them, is to be driven down in the water and firmly propped there. Then the lower surface inside, under the water, must be laid across, and finally concrete from the mortar trough must be heaped up until the empty space which was within the cofferdam is filled up by the wall.'

After the collapse of the Roman Empire the art of bridge building

The 7th-century Great Stone Bridge in China. To reduce the weight of the structure the spandrels were pierced. This form of construction was seldom used in Europe before the mid-18th century.

The first stone bridge over the River Thames in London, completed in 1209. It was a relatively crude structure compared with others built in the early medieval period.

in Europe fell into decline. Elsewhere some remarkable bridges were built, in particular the Great Stone Bridge in China. This slender arch of 123 foot span and 23 foot rise from the abutments to the crown was built between 589 and 618. Compared with the Roman semicircular arches, this is an extremely shallow structure for a stone arch, producing a large thrust at the abutments. To reduce the weight of the structure the spandrels were made open, while retaining a relatively shallow deck. In Europe before the mid-18th century there were few examples of this form of construction, which is similar to the elegant reinforced-concrete arches constructed by Robert Maillart at the beginning of the 20th century.

Most of the road and bridgeworks carried out in early medieval days were directed by the monasteries. The first stone bridge to be built across the Thames at London (the Romans and Saxons built timber bridges) was begun in 1176, and the work was directed by a

priest, Peter Colechurch. The structure consisted of 20 arches and was completed in 1209, four years after Colechurch's death. The pier foundations were built up from timber piles and a row of timber houses was built along the deck. London Bridge was a relatively crude structure compared with others built during the medieval period like the Pont St Bénézet at Avignon begun in 1177, part of which remains intact today. Most medieval bridges incorporated a chapel or fortifications. Narrowed roadways were also common and made defence of the structure easier. Other features of medieval bridges were the use of triangular projections from the piers, known as cutwaters, and the development of the ribbed arch. The cutwaters streamlined the flow of water round the piers and at the same time formed a recess at road level for pedestrians to shelter from passing carts and horsemen. The use of separate arch ribs, across which thinner stones could be laid, reduced

The celebrated Pont St. Bénézet at Avignon, begun in 1177. It is one of the finest examples of early medieval arch construction.

29

Three medieval bridges: Monnow Bridge (note the use of separate arch ribs); Chapel Bridge, Wakefield (opposite); and Newby Bridge (below opposite). The flow of water was streamlined around the piers by cutwaters.

Fig. 33. Ribbed construction, reducing the quantity of material in the deck, is extensively used in modern steel and concrete bridges.

the quantity of material required and so cut down the load on the foundations. Ribbed construction is extensively used for modern steel and concrete bridges for the same reason (figure 33).

Covered bridges in stone and timber were also built in this period; the most famous example is the Ponte Vecchio at Florence, erected in 1345. An unusual feature of this bridge is the three segmental arches with spans of up to 100 feet. This was a departure from earlier bridges in which the arches consisted of two or three centred curves. There are shops on both sides of the bridge, above which is a gallery connecting the Pitti and Uffizi palaces.

The Renaissance saw the evolution of the scientific attitude to the design of structures: Leonardo da Vinci and Galileo Galilei made several important advances in the understanding of the strength of materials though little notice was taken of their work at the time.

Leonardo designed a light bridge structure, capable of easy transport, which had obvious military applications. Considerable development of the truss principle was made during this period. Palladio, the famous Italian architect, worked out several designs for wooden trusses, which were used for spans up to 100 feet. The arch bridge is perhaps the most characteristic of Renaissance engineering skill— segmental arch curves replacing the semicircles used by the Romans. Two examples that illustrate the beauty of Renaissance arch bridges are the Rialto Bridge (completed in 1591) in Venice—a segmental curve of 21 foot rise and 89 foot span—and the Ponte San Trinita, Florence (completed in 1569). Here the 'basket-handled' curve is

Timber trusses designed by Palladio, which were used for spans up to 100 feet.

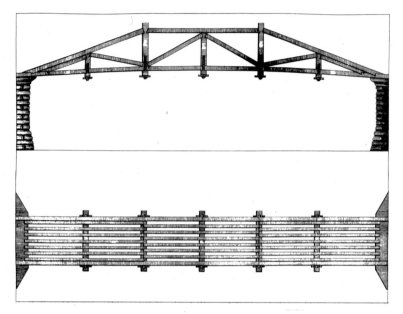

Below, the shorter arm of Pont Neuf, Paris, with five arch spans ranging from 33 to 52 feet.

34

The longer arm of Pont Neuf (reconstructed in 1850), with seven arch spans ranging from 52 to 62 feet.

formed by parts of two parabolic arcs, which meet at the crown in an obtuse angle concealed by an escutcheon. The Italian architect-engineers of the Renaissance influenced work in France, of which the famous Pont Neuf connecting the Ile de la Cité with the Left and Right Banks of the Seine is a notable example. The first pier was laid down in 1578 but work was interrupted by civil war and the structure was not finally completed until 1604. The bridge consisted of two series of masonry arches connecting the island; the overall span was about 770 feet and the width 65 feet. The arch spans were 52 to 62 feet for the larger arm (seven arches) and 33 to 52 feet (five arches) for the shorter arm. The bridge was reconstructed in 1850 to improve the alignment of the carriageway.

During the 18th century scientific methods were gradually introduced into various fields of engineering. Mathematics and structural theory were rapidly developing and in 1747 the famous Ecole des Ponts et Chaussées was founded in Paris for training engineers in road and bridge works. At this time France was ahead of other countries in the development of structural theory and the Ecole played an important part in the evolution of modern bridge engineering. The school's first director was Jean-Rodolphe Perronet (1708–94), one of the world's greatest designers of arch bridges. Two of his best works were the Pont Neuilly, crossing the Seine below Paris, and the Pont de la Concorde in Paris. The Neuilly Bridge had five arches of about 120-foot span with a 30-foot rise. The interior of the arches was made up of a series of tangential circular arcs and the exterior was segmental in form, the difference in curvature being taken up by a bevelled inward slope. Perronet appreciated that for a series of equal arch spans the thrusts at the intermediate piers would be balanced for the dead load of the bridge; he thus adopted a small pier width to span ratio, which increased the waterway opening and improved the appearance of his arches. This ration for the Neuilly Bridge was about 1:10. At this time it was common practice to use a ratio of 1:5 while the Romans generally adopted a ratio of 1:3, each pier acting as an abutment so that if one span was destroyed the others would not collapse. It is essential that all the arches are built before the centering is removed if the principle of interaction is to be followed. Peronnet also devised a new form of pier construction consisting of two columns connected by a lateral arch, which reduced the volume of stonework. Perronet hoped to make the Pont de la Concorde his finest work; but he was hampered by officialdom and forced to make the piers solid and increase the rise of the arches by three feet to improve the headroom for river traffic. The bridge was completed in 1791 and consisted of five masonry arches with spans of 81, 91, 101, 91 and 81 feet. The width of the bridge was increased from 45 feet to 114 feet in 1931. In the latter half of the 18th century the design of masonry arches reached an extremely high standard, which remained unsurpassed in the 19th with a few exceptions, including John Rennie's Waterloo Bridge (nine semi-elliptical arches of 120 feet) completed in 1817.

The completion in 1779 of the cast-iron arch bridge spanning the River Severn at Coalbrookdale in Shropshire marked the end of the reign of stone and timber as the dominant structural materials for bridge construction, although the details of the Iron Bridge show that it was built in the spirit of timber and masonry. Cast iron is unreliable in tension and is most appropriate for arch bridges. The Iron Bridge aroused interest and led to the extensive use of metal for 19th-century bridge works. At the beginning of the 19th century, wrought-iron chains were being used for suspension bridges and the pioneer work in this form of construction was that patented by James Finley in 1808. The most ambitious of Finley's works was the footbridge across the Schuylkill River at Philadelphia completed in 1809 with a 308-foot span. The finest example of early suspension-

The completion of the cast-iron bridge near Coalbrookdale in 1779 marked the end of the reign of stone and timber as the dominant materials for bridge construction.

bridge design was Thomas Telford's Menai Bridge of 579-foot span completed in 1826.

There were many failures of early suspension bridges, which often had light timber decks; not surprisingly they were thought to be unsuitable for heavy railway loadings. Railway engineers therefore used different structural forms for their bridges and cast iron was gradually replaced by wrought iron because of its greater reliability in tension. The first iron railway bridge in Britain carried the Stockton and Darlington railway over a small stream, the Gaunless at West Auckland. The bridge was replaced in 1901, but the original structure is permanently on display in York Railway Museum. The five spans consist of fish-belly girders, the curved members being wrought-iron bars $2\frac{1}{4}$ inches in diameter. This bridge embodies an interesting structural concept—the combination of the arch and chain (figure 34). The outward thrust of the top compression boom is balanced by the inward pull of the bottom tension boom. The bridge was opened in 1825; the designer was George Stephenson and the contractors John and Isaac Burnell.

The most significant bridge structure of the 19th century was

Above, model of the first iron railway bridge, built for the Stockton and Darlington Railway in 1824.

Right, the Garabit Viaduct designed by Eiffel and completed in 1888. The structure was designed to resist the very strong winds which blow along the gorge.

Fig. 34. The structural principle of the Gaunless Bridge combined the arch and chain.

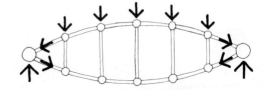

Robert Stephenson's Britannia tubular bridge completed in 1850 for the Chester and Holyhead railway. The structural concept of this bridge is of great interest to engineers throughout the world. The bridge was originally conceived as a series of large tubes (through which the trains could pass) supported by chains. However, model tests and theoretical work carried out by William Fairbairn and Eaton Hodgkinson indicated that the chains could be discarded and

40

The Britannia Tubular Bridge, completed in 1850 for the Chester and Holyhead Railway. The experimental work carried out in the design of this bridge is still of great interest to engineers.

Right, Glenfinnan Viaduct, Inverness-shire, a multi-span arch structure completed in 1898. It was the first railway bridge in Great Britain to use concrete except as an auxiliary material.

The reinforced-concrete bridge over the Rhine at Tavanasa, Switzerland, designed by Robert Maillart and built in 1905.

A prestressed-concrete bridge over the Rhine at Worms, Germany, built by the cantilever method of construction.

Many large suspension bridges were built in North America during the 1930s, including the Golden Gate Bridge, San Francisco, span 4200 feet.

One of the most famous steel arch bridges – Sydney Harbour Bridge, Australia.

that the tubes would be strong enough to carry loads without additional support. The basis on which the design was evolved using theory and experiment is typical of the approach adopted for present-day bridge structures. Important developments in other parts of the world during the latter half of the 19th century included the Squire Whipple truss bridges, Roebling's suspension bridges in America and Gustave Eiffel's truss-and-arch viaducts for the French railway network. The completion of the bridge spanning the Firth of Forth at Queensferry in 1890, with its powerful cantilever arms and main span of 1700 feet expressed the tremendous achievements of the 19th-century engineers, and the immense strides made in the technique of bridge design and construction since the first iron bridge over the River Severn was completed some 100 years before.

In the late 19th century, concrete began to be used extensively for bridge construction. Between 1887 and 1891, Wayss and Freitag of Frankfurt-am-Main constructed over 300 concrete arch bridges with spans up to 130 feet. The first use of concrete in Great Britain for a railway bridge, other than as an auxiliary material, was in 1898 for the Glenfinnan Viaduct.

The limited resistance of concrete to tension led to the develop-

ment of reinforced concrete in which steel was introduced to resist the tension. The theory of reinforced concrete was largely developed by European engineers, notably François Hennebique, who used a composite material (steel and concrete) for the design of bridges. The most interesting early reinforced concrete bridges were the concrete arches built by Robert Maillart. These bridges illustrate

A view of bridges spanning the Thames in London: Westminster, Charing Cross, Waterloo, and Blackfriars.

Right, a few of the many elegant bridges spanning the Seine in Paris.

44

how simplicity of line and economy of material are combined to produce an elegant structural form.

The most significant structural advance of the 20th century was the technique of prestressing concrete. An initial compressive stress is induced in a structure to eliminate tension caused by the loading, the tension causing the concrete to crack. The technique of prestressing was largely developed by Eugène Freyssinet but was not

The underside of the Hammersmith Flyover, showing details of the prestressed-concrete construction.

Opposite, the James Lich Freeway, San Francisco.

used extensively for bridge works until the late 1930s. It is now the dominant material for bridge construction. The construction of the first prestressed concrete bridge over the River Rhine at Worms, in 1952, by a cantilever method of construction aroused considerable interest as one of the most impressive achievements in prestressed concrete construction.

With the decline of the railways the monumental bridges of the 20th century have been built mainly for motor traffic. Many major cities have large and famous bridges—the Golden Gate Bridge in San Francisco, Sydney Harbour Bridge, Verrazano Narrows

Construction of the Mancunian Way, an elevated dual-carriageway with precast box sections stressed together to form a continuous girder.

Right, Bonn/Nord Bridge, Germany. The steel deck girders are braced by cables to reduce their effective span.

48

Polcevara Creek Bridge, Genoa. The concrete box girders are braced by cables.

Bridge in New York, the Thames bridges in London and the Seine bridges in Paris to name only a few. Many of the older bridges in cities are now becoming inadequate for modern vehicular loads and traffic flow. In Paris several famous bridges have been strengthened or replaced. Meantime the building of elevated urban roads and motorways has made it necessary to construct thousands of small- and medium-span bridges, in which the dominant material is concrete, reinforced or prestressed. Wide use has been made of model tests in the design of elevated roads such as the Hammersmith Flyover and the Mancunian Way. An important development in the past 15 years has been the use of welded-steel box girders braced by cables to reduce their effective span. To a lesser extent the same system has been used with hollow prestressed concrete girders.

In the latter part of the 20th century bridge design may well dominate construction of transport systems. Monorail systems are now being considered for many city centres; these will involve engineers in even more sophisticated structural systems and materials.

The rapid development of motorways in the present and of motor-

rail systems in the future means that time is an increasingly important factor in the overall economics of bridge construction. Together with the tendency towards minimum-weight design, the need for time-saving has led to the development of designs and methods of construction that depend on the use of electronic computation. The use of a computer has relieved the engineer of the mass of numerical manipulation involved in modern bridge design, which can too easily cloud the fundamental considerations of safety and economics. Obviously the computer can never replace the engineer, but if used correctly, it enables him to select on a more rational basis the correct bridge structure for a specific situation. The entirely automatic design of certain types of bridge has already been developed and the automatic production of detailed drawings and quantities is imminent. However, it must not be supposed that the computer is now able to solve all the problems confronting the bridge engineer. It can do no more than operate on a set of instructions. It is the quality of the information that goes into the computer that is all-important. And as our civilisation develops, bridge construction will present an even greater challenge to man's ingenuity.

A concrete monorail structure in Japan which runs from Tokyo to Haneda Airport.

Iron Bridge at Coalbrookdale

The completion in 1779 of the Iron Bridge spanning the River Severn near Coalbrookdale in Shropshire marked the end of the reign of stone and timber as the dominant materials for bridge construction. Its 100-foot span is modest by 19th and 20th century standards; yet it is generally acknowledged to be the first large-scale bridge to be built completely in cast iron. Not until the end of the 18th century was iron employed as a major structural material, particularly to replace timber columns in industrial buildings, although wrought-iron chains were used for suspension bridges in China and India as early as the first century AD.

Iron as a free metal is rarely found on the earth's crust. It occurs however in many minerals near the surface. The metal-bearing minerals are referred to as ores, from which the iron is extracted by smelting with carbon. Until the end of the 16th century, wood charcoal was used for smelting and timber was in such demand that the supply for shipbuilding was affected. Then, in the first half of the 17th century, Dud Dudley (an illegitimate son of the Earl of Leicester) realised that iron could be smelted more efficiently by replacing wood with coal. Even so, the potential of this method was never realised and it was not until the early 18th century that the use of coke for smelting iron was developed by Abraham Darby. He started a smelting business at Coalbrookdale, in a district that had been the centre of the iron industry since Tudor times. Abraham Darby died in 1717, and his son of the same name grew up to continue his father's trade of ironfounding. Abraham Darby the second was succeeded in the management by his son-in-law Richard Reynolds in 1763. Thus the first iron bridge was built in a district with a great tradition of ironfounding.

As we have seen, at the time the Iron Bridge was built, stone was still the dominant structural material and many fine masonry bridges were constructed in Shropshire and in Wales. For example, William Edwards (born in 1719) constructed several flat arch bridges in stone, including the Dolau-Herion bridge over the River Towy near Llandovery, Wales and the Pontypridd bridge over the River Taff, completed in 1755. To reduce the self-weight of these structures, the spandrels were pierced by holes; this pointed the way to the use of the open-spandrel arch to minimise the self-weight. Nearer to the

A masonry bridge built by William Edwards over the River Towy, near Llandovery, completed in 1740. The pierced spandrels are almost completely hidden by foliage.

site of the Iron Bridge, John Gwynn (died 1786) constructed some elegant stone bridges over the River Severn including the Atcham Bridge and the English Bridge at Shrewsbury (1769–74). The significance of the Iron Bridge itself might well have passed unnoticed had not Thomas Telford been appointed Surveyor of Public Works for the County of Shropshire in 1787, at the age of 30. Telford had been trained as a stonemason but was soon to become the first and greatest master of cast iron. He went to Shrewsbury, some 14 miles from Coalbrookdale, to supervise the renovation of the castle, and his first bridge was at the nearby village of Montford. It was of arch form, built of red sandstone and completed in 1792. Indeed in the years 1790–96 Telford built many stone bridges in Shropshire but also, more important, a cast-iron bridge at Buildwas, a short distance from Coalbrookdale. The Buildwas Bridge was a flat arch with a span of 130 feet, 30 feet longer than the Iron Bridge, and, more significantly, half the weight. It is interesting to speculate how much Telford's adoption of cast iron to replace stone was influenced by the Iron Bridge.

The idea of building the bridge completely in iron was at one time attributed largely to Abraham Darby, but recently the situation has been reassessed by Peter Matthews and Robert Maguire while

Pontypridd Bridge over the River Taff, completed in 1755. The spandrels are pierced by three circular holes on each side of the bridge to reduce self-weight.

Atcham Bridge over the Severn, built by John Gwynne.

English Bridge, Shrewsbury, built by John Gwynne and completed in 1774.

Montford Bridge over the Severn. This was one of the first bridges built by Thomas Telford in Shropshire and was completed in 1792.

studying at the Architectural Association in London. To summarise their findings: several people were in fact involved in the design of the Iron Bridge, including Thomas Pritchard, a Shrewsbury architect and two rival iron-masters, Abraham Darby and John Wilkinson. The part played by Pritchard in the conception of an iron bridge was not appreciated until Matthews and Maguire uncovered further information in the Science Museum Library. They discovered that Pritchard, who, it is generally agreed, played no part in the final design, conceived the idea of using iron for bridges some years previously.

The first meeting to discuss the possibility of constructing a bridge over the River Severn near Coalbrookdale was held in September 1775. It was agreed that Pritchard and one Samuel Thomas were to prepare estimates. Neither Pritchard nor Darby were present, though the meeting was attended by Wilkinson. Earlier, Pritchard had prepared a design for a bridge, illustrated on p. 57, which basically was a masonry structure on a cast-iron centre. The spandrels were pierced by holes in a similar manner to the Dolau-Herion and Pontypridd bridges constructed in masonry some 20 years previously.

At a further meeting, held in October 1775, Pritchard presented another design for a bridge to be constructed completely in iron and Abraham Darby was commissioned to build it. The structure was a simple arch of 120-foot span with the centre 35 feet above low water level and the deck 18 feet wide. In the following year, several meetings were held but the dimensions of the structure remained unchanged. However, at a meeting held in July, 1777, a third bridge, of different dimensions, was being discussed. 'Mr Abraham Darby agrees to erect an Iron Bridge in a substantial manner, the superstructure not less than 24 feet wide and an arch over the river 90 feet span with a proper towing path under the same, any necessary alterations to be agreed upon at a future meeting.'

From this description it is apparent that Pritchard's second design would not fit in with the specification. The road had been widened by 6 feet and the headroom allowed by Pritchard was inadequate for the sailing barges used on the River Severn at that time. Pritchard died three months after the July meeting, and the responsibility for the design of the completed structure with its 100-foot span still remains in doubt, although it was built by Abraham Darby.

The following description of the construction of the Iron Bridge

Pritchard's design for a bridge over the Severn near Coalbrookdale. The first scheme employed a masonry arch with pierced spandrels on a cast-iron centre, and the second an arch constructed completely in cast iron.

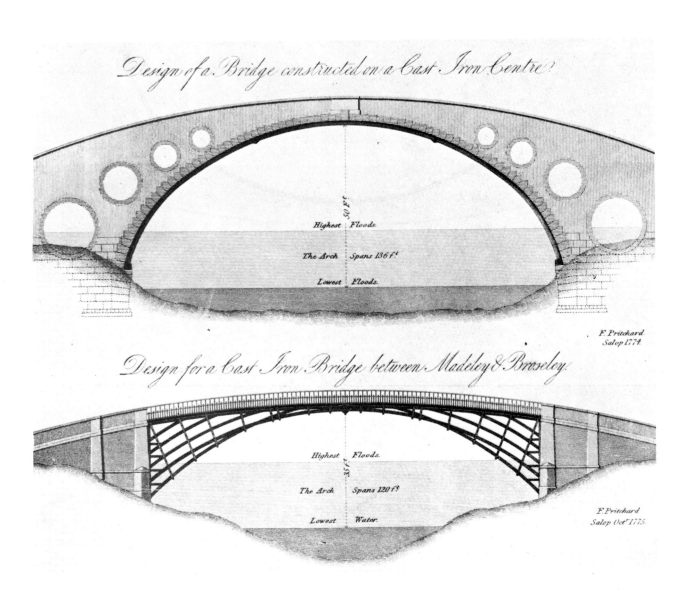

Design of a Bridge constructed on a Cast Iron Centre.

50 Ft.

Highest Floods.

The Arch Spans 136 ft.

Lowest Floods.

F. Pritchard.
Salop 1774.

Design for a Cast Iron Bridge between Madeley & Broseley.

35 Ft.

Highest Floods.

The Arch Spans 120 ft.

Lowest Water.

F. Pritchard
Salop Oct.r 1775.

is taken from an account published in an engineering magazine: 'The span of the arch is 100 feet 6 inches and the height from the base line to the underside of the arch at the apex is 40 feet. The total weight of iron used in the making of this structure amounted to 378 tons 10 cwt, the castings being made at the yard of the Coalbrookdale Company. The bridge was necessarily of cold-blast iron, as no

The shorter ribs pass through apertures in the vertical pillars.

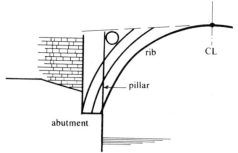

A line drawing representing one set of ribs for the bridge.

rib

CL

pillar

abutment

other was then made, and it is surmised that the members were cast from a re-melting in an air furnace. The weight-carrying ribs and braces were open castings, only the balustrade being covered castings. The five main ribs were cast in two parts only, approximate length 70 feet, and of plain rectangular section 9 inches by 7 inches and 9 inches by 6 inches solid. It is doubtful whether any foundry in England at the present time could, or would, undertake to make castings such as these. Each of the large ribs weighs 5 tons 15 cwt.

The floor 24 feet wide is of open sand cast-iron plates 3 feet 8 inches wide, each plate 24 feet by 3 feet 8 inches by $2\frac{1}{2}$ inches flat, not ribbed. The bridge was made and erected in the year 1779. For erection a large scaffold was built clear of the river traffic, each part of the ribs was lifted to a height with ropes and chains and then lowered until the ends met at the centre, and there fastened with a dovetail joint and iron keys and screws—no bolts were used. The screws were made in wrought iron, cut with a thread designed by Abraham Darby (at that time no standard of screwing was known). The main parts of the structure were erected in three months

The shorter (inner) ribs are mortised into the top bearers and into the base plate and pillar.

without delay to river traffic (then so very important, as iron and iron goods from Coalbrookdale and Horsehay Works were conveyed by barges to Bristol and other towns—it has to be remembered that there were no railways in those days). It is recorded that there was no accident to the bridge itself, or to the workmen, during the whole of the erection.

The line drawing (on p. 58) represents one set of ribs of the bridge. On the abutments of stonework are placed iron base plates with mortices, in which stand the pillars. Against the foot of the inner pillar the bottom of the main rib bears upon the base plate. The rib consists of two pieces and is connected by a dovetail joint in an iron key and fastened with screws; each piece is 70 feet long. The shorter ribs pass through the pillars at apertures left for that purpose and are mortised into the top bearers and into the base plate and pillar; the cross stays, the braces, the circle and the brackets connect the larger pieces to each other, so as to keep the bridge steady. The diagonal stay, the cross stays and the top plates answer the same purpose by connecting the pillars and ribs to each other in the

The cast-iron circle recalls the traditions of masonry (pierced spandrels) in which the bridge was built.

Opposite, the Iron Bridge. This pioneering structure, the first bridge in the world to use iron as the basic construction material, is today in danger of collapse.

opposite direction. The whole bridge is covered with iron top plates projecting over the ribs on each side. On this projection stands the balustrade of cast iron. The road over the bridge made of clay and iron slag is 24 feet wide.

The overall dimensions of the bridge are those of a masonry arch of the same period and the cast-iron circles are comparable with the pierced spandrels of William Edwards' masonry structures. The detailing of the members is similar to that of high-quality timber construction and the members were cast in such a way that they could be assembled by slotting some members through others at various points. The holes were made large enough to allow for the considerable swing and movement that occurred during positioning and this play was taken up by cast-iron wedges on completion. The joints are either held in position by wedges or else each member drops into a dovetail box on another and is tightened up with a screw.

The bridge is now closed to all vehicles except perambulators and pedal cycles pushed by hand, and a notice on the handrailing reads as follows: 'This iron bridge was erected in 1779 and was then the first cast-iron bridge. It was made at Coalbrookdale by Abraham

The members were assembled by slotting, and the play was taken up by cast-iron wedges in a manner similar to that used in timber construction.

Left, diagonal and cross stays were used to increase the stability of the slender arch ribs.

Darby to the designs of Thomas Farnoll Pritchard of Shrewsbury. It was closed to vehicular traffic in 1934 and in the same year was scheduled as an ancient monument. The proprietors of the iron bridge handed it over to the Shropshire County Council on 12th October 1950.'

Notwithstanding the slightly derelict appearance of the Iron Bridge, it is difficult to understand how it can be regarded by some locals as a monstrosity. It has the rare quality of being picturesque and at the same time conveying, after a brief inspection, real engineering qualities in its conception and construction details. It must be remembered that although the structure was not calculated, considerable engineering skill was required for the casting, erection and design of the joints. Also, at this time, rivets and bolts had not been invented. The outer ribs have a specially moulded rib below which is lettering stating: 'This bridge was cast at Coalbrookdale and erected in the year MDCCLXXIX'.

Yet in spite of its great historical significance the bridge has not been well maintained. Chicken wire has been draped between the cast-iron members some eight feet above the base plate on the north bank. Its purpose is to collect loose material breaking away from the underside of the deck and thus prevent it from falling on to people using the tow-path. On the north bank, the bridge approaches are used as a bus stop. On the south bank there is a disused railway line, the windows of the toll cottage are bricked up, and there is speculation over how long the structure will remain intact. It is suggested that the south bank is gradually slipping, with the result that an increasingly pronounced kink is developing at the crown of the arch. Someone standing on the south bank can see a distinct kink at the crown of the bridge, with the first-floor windows of the Tontine Hotel, which faces the bridge, just visible above it. It is claimed that at one time the top of the door at ground level was visible from the same position.

Indeed not only the bridge but the whole area of Coalbrookdale gives an impression of stagnation and decay. What was once a thriving centre of the iron industry is now a tranquil stretch of the River Severn. The bridge itself is surrounded by vegetation, in contrast to the almost clinical appearance of the Albert Edward railway bridge, cast and erected by the Coalbrookdale company in 1863. This bridge is sited about one mile upstream towards Shrewsbury.

Clifton Suspension Bridge

It is difficult to imagine a more romantic setting for a bridge than the Avon Gorge at Clifton, Bristol. The suspension bridge that spans this magnificent gorge was the product of the engineering genius of Isambard Kingdom Brunel. Brunel was commissioned to construct the bridge in 1830, at the age of 24, as the result of a competition. The project was completed in 1864, some years after his premature death. It was preceded by several notable suspension bridges and before examining Brunel's great work it might be useful to describe briefly the development of this type of bridge up to 1830.

As we saw in the Introduction, there are records of chain bridges in China and Kashmir built as early as the first century AD. The first recorded appearance in Europe was the publication of a drawing of a suspension bridge by Fausto Veranzio in 1595. In 1741 an iron chain footbridge was constructed across the River Wear in Durham, England; this was probably the first suspension bridge to be built in Europe, but no further examples were constructed until the early part of the 19th century. The first iron chain suspension bridges which were capable of carrying vehicular traffic were developed by James Finley in America. He invented a flexible chain of wrought-iron links, basically similar to the Durham bridge, and used wooden trusses and transverse wooden beams for the deck. The basic problem in the construction of these bridges was to determine how long the cable and each hanger should be in order to obtain a level deck. Finley's description of how this should be tackled is as follows:

'To find the proportions of the several parts of a bridge of one hundred and fifty feet span, set off on a board fence or partition one hundred and fifty inches for the length of the bridge, [then] draw a horizontal line between these points representing the underside of the lowest tier of joists. On this line mark off the spaces for the number of joists intended in the lower tier, and raise perpendiculars from each, and from the two extreme points; then fasten the ends of a strong thread at these two perpendiculars, $23\frac{1}{4}$ inches above the horizontal line. The thread must be so slack that when loaded, the middle of it will sink to the horizontal line; then attach equal weights to the thread at each of the perpendiculars, and mark carefully where the line intersects each of them.'

1859 died, 53 years old.

Finley received a patent for his invention in 1808, and between 1808 and 1816 about 40 chain bridges were built. The largest span that Finley attempted was a footbridge across the Schuylkill River at Philadelphia in 1809 with a 308-foot span between the towers.

Meanwhile in England Thomas Telford was consulted in 1814 about a scheme for bridging the River Mersey at Runcorn. With the help of Peter Barlow and others, he carried out a series of tests on malleable iron; these resulted in a proposal for a suspension bridge with a central span of 1000 feet and two side spans of 500 feet. (It is possible that Telford knew something of Finley's work and this may have influenced his proposals.) The project was later abandoned, but in 1818 the Holyhead Road Commissioners asked Telford to submit plans for bridging the Menai. Telford's design provided for a single span of 579 feet between piers, which were approached by masonry arches. The bridge was opened to traffic in 1826. At this time English engineers had little interest in structural theory; Telford disliked mathematics and was hardly acquainted with the elements of geometry. He did, however, appreciate the importance of tests and found that cables stretched rapidly at a force of little more than half the breaking value. It was thus decided that the cables should not be subjected to loads greater than one third of the breaking value.

In keeping with the tradition of the engineering schools, the theory of suspension bridges was developed in France. The engineer Claude Navier was sent by the French government to England to study the art of building suspension bridges in 1821. He was very impressed with the Menai Bridge, which was then under construction, and in 1823 published a report which included a description of existing bridges and theoretical methods of analysing them.

The designer of the Clifton Bridge, Isambard Kingdom Brunel, was born in 1806. He was the only son of the French-born engineer Marc Brunel, and completed his education in France. Brunel began his engineering career by working for his father on the Thames Tunnel, where he narrowly escaped death. He was sent to Clifton to convalesce, and there he heard of the proposal to span the Avon Gorge and to invite engineers to submit designs.

The Avon Gorge, the sides of which rise over 300 feet above river level, is about five miles from the mouth of the Avon. The first known design for spanning the gorge was produced by W. Bridges; this

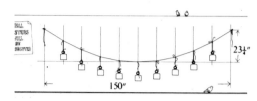

A method suggested by James Finley for establishing the length of the cables and hangers for a suspension bridge.

was a stone construction filling most of the gorge. Before submitting his designs Brunel examined existing suspension bridges, including Telford's Menai Bridge. He submitted four designs with spans varying from 860 to 916 feet. The bridge committee asked Thomas Telford to examine the designs and he rejected them all. At this stage in his career Telford had become concerned about the stability of long-span suspension bridges when subjected to wind pressures and was very dubious of Brunel's proposals. He was also over 70, and his decision to reject Brunel's proposals may have been influenced by the conservatism of old age. Consequently Telford was asked to submit his own design and produced a three-span structure with colossal Gothic piers. Popular opinion was not, however, in favour of Telford's design, and it was decided to hold a second competition. Twelve designs were submitted, including Telford's, of which four were selected for final assessment, excluding Telford's and including Brunel's. It had been specified that the intensity of load on the suspension chains should not exceed five and a half tons per square inch and only Brunel's design fulfilled this requirement.

At this time the first wire-rope suspension bridge had been constructed by French engineers. When wire is drawn, its tensile strength is increased. And if a large number of wires are compacted into a single unit, the resulting mass will have a much greater tensile strength than a solid bar of equal cross-sectional area. After a visit to the continent, Mr West of the Clifton Observatory was asked by

Menai Suspension Bridge designed by Thomas Telford, completed in 1826 and still in service. It has a span between towers of 579 feet, which was the world's longest until the construction of the Fribourg Suspension Bridge in Switzerland, span 870 feet, in 1834.

67

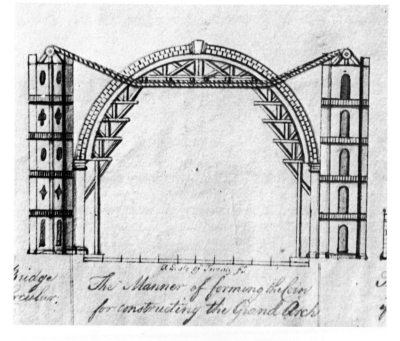

W. Bridges' design for a bridge spanning the
Avon at Bristol. Above, the complete structure
with an arch width of 140 feet. Right, the
method proposed for constructing the arch rib
on timber staging.

the bridge trustees to prepare a design for a less expensive bridge utilising wire cables. This alternative design included less ornate and smaller towers and a lighter deck. (A feature of Brunel's bridge was the Egyptian design for the towers, which contrasted with the slender chains and suspension rods.) However, these amendments produced much dissatisfaction and Brunel's original design was retained.

The first stone for the bridge was excavated on the Clifton side of the gorge in 1831. In this year, however, there were riots in Bristol and little work was carried out for a few years. Before the laying of the foundation to the Leigh abutment in August 1836, an attempt was made to take an iron bar, $1\frac{1}{2}$ inches in diameter, across the gorge. Eventually the bar was placed across the gorge but not

One of Brunel's designs for the Clifton Bridge. The Egyptian appearance of the towers contrasted oddly with the slender chains and suspension rods.

without developing a bend. In September 1836, a new bar was placed in position, and Brunel travelled across in a basket. In the next twenty years work progressed very slowly because of financial difficulties and at the time of Brunel's death in 1859 the two piers had not been completed.

Shortly after Brunel's death, a company for the completion of the bridge was formed by members of the Institution of Civil Engineers as a tribute to Brunel. The demolition of Hungerford Suspension Bridge over the Thames in London (also built by Brunel between 1841 and 1845) provided about 70 per cent of the necessary iron-work, which was acquired at a very low price. An Act of Parliament authorising the construction of the bridge was obtained in 1861, and J. Hawkshaw and W. H. Barlow were appointed engineers. Hawkshaw's and Barlow's scheme was somewhat different from Brunel's original design: the width was increased from 24 to 30 feet and the height from 230 to 245 feet. To use the Hungerford Bridge chains in the most economical way, three were placed each side of the bridge instead of the two proposed by Brunel. The deck girders were entirely of iron instead of a combination of wood and iron and the anchorages were brought nearer to the piers.

The Hungerford Bridge chains became available at the end of 1862 and the first step was to place the chains across the gorge. This was achieved by placing temporary wire-rope staging between the two piers. Laying the chains was completed in May 1864. The load from the chains is transmitted to the piers via wrought-iron saddles; the saddles rest on cast-iron roller frames incorporating cast-steel rollers. These rollers allow the saddles to move in either direction as the chains expand or contract. The rollers, 2 feet long and 4½ inches in diameter, are located between the main saddles and a cast-iron base. Two of the three chains are attached to the main saddle and the third to a secondary saddle, which is bolted to the main saddle. As two sets of chains pass over each pier four saddles in all are required—two to each pier. When the chains reach ground level they are deflected down into tunnels leading to the anchorages by means of fixed wrought-iron saddles. The tunnels are cut through solid rock and the anchorages are 70 feet below ground level. Access to the anchorages was provided by a shaft located outside the toll houses.

The next stage was to attach the suspension rods, which are at 8-foot intervals along the length of the bridge, to the chains. At the same time the construction of the main longitudinal girders was

Passengers being carried across the gorge in a basket supported by a single bar.

Right, three sets of chains were used on each side of the bridge, though Brunel's original proposal was for two sets.

Work in progress on the Clifton Bridge. The chains were assembled link by link on a timber platform, supported by wire ropes, between the towers.

The view in 1864 looking down from the tower on the Clifton side. The cross girders were bolted to the underside of the longitudinal girders and cantilever each side to support the decking to the footways.

The suspension rods were attached to the chains at 8-foot intervals along the length of the bridge. At the same time, the 3-foot-deep main longitudinal girders were erected in 16-foot sections.

begun. These girders were placed in 16-foot sections and are 3-foot deep. The girders separate the carriageway from the footpaths and provide the principal stiffening to the structure. The cross-girders were then bolted to the underside of the longitudinal girders together with cross-bracing. The carriageway floor consisted of 5-inch thick timbers, tongued and grooved together, on top of which were placed 2-inch thick floor planks.

construction.

Before the bridge was officially opened on December 8, 1864, it was subjected to a test load of 500 tons. A deflection of 7 inches was observed at the centre of the bridge and the saddles moved forward 1½ inches. All suspension bridges are liable to move because of wind loading, and when crossing the bridge in a high wind one feels rather as if one is on the deck of a ship in a heavy swell. Indeed, movements of more than 6 inches above and below the normal carriageway level have been observed.

The Clifton Bridge has been carefully maintained since it was

Maintenance work is carried out from a cradle slung from the chains.

Right, repainting in progress. The bridge has been carefully maintained since its completion in 1864.

built, and in 1887 the timber deck was surfaced with asphalt. Maintenance work on the underside of the bridge has been and still is carried out from a cradle suspended from the handrails. Other work during the past 80 years has included strengthening both anchorages, replacing the maintenance cradle, strengthening the chain links next to the land saddles, and renewing all the timbers and bolts securing the cross-girders. In 1955 it was decided to strip the cross-beams down to bare metal because of the increasing air pollution. The metal was then zinc-sprayed to give lasting protection from atmospheric corrosion.

conclusion.

The careful preservation of the Clifton Bridge is a fitting tribute to the engineering genius of Brunel. The bridge is a constant source of inspiration to engineers and it is amazing that the design was conceived by a man of only 24; when first seen, this superb structure gives the bridge designer a sense of pride and humility.

Britannia Tubular Bridge

Thomas Telford's suspension bridge over the Menai Straits was completed in 1826. It carried the London to Holyhead road from the mainland to Anglesey and was an important development in the history of bridge engineering. With the coming of the railway age, a bridge of even greater significance—and of different structural form—was built to carry the Chester and Holyhead railway. This was the Britannia Tubular Bridge, opened in 1850, 24 years after the completion of Telford's suspension bridge. It was a major landmark in the development of bridge and structural engineering. Indeed the techniques used are similar to those used today, not only for bridges but also for aircraft and buildings.

The basic structural form is a tube, rectangular in section, through which the trains pass. It is supported by five piers and is the first notable example of a wrought-iron continuous beam structure. The Britannia Bridge has an international reputation even today, over a century after its completion, and the experiments carried out in connection with its construction are of great interest to engineers working on thin-walled structures. In the design of modern bridges it is becoming more and more important to aim for minimum self-weight. Structures with wasteful proportions are almost always uneconomic.

Tubular sections are now common for short-, medium- and long-span structures of both steel and concrete. Their design often involves the testing of models to ensure a complete understanding of their structural action. But as we have already seen, structural theory based on combined mathematical and experimental investigations was practically unknown until the 17th and 18th centuries. During the early part of the 19th century, the mathematical theory of structures was largely developed by French engineers, and in England experimental work provided a background for the solution of important structural engineering problems.

The designer of the Britannia Bridge was Robert Stephenson (1803–59), the only son of George Stephenson (1781–1848) of 'Rocket' fame, known as the 'Father of Railways'. After attending classes in science at the University of Edinburgh, Robert Stephenson assisted his father in surveying the Stockton and Darlington and Liverpool and Manchester railways. On the first both steam

Britannia Tubular Bridge, an early example of fruitful co-operation between engineer and architect, was opened in March 1850.

76

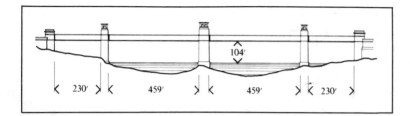

The Britannia Bridge was the first notable example of a wrought-iron continuous-beam structure. The tubes through which the trains pass are rectangular in section and continuous over four spans. The main dimensions of the bridge are shown in this sketch.

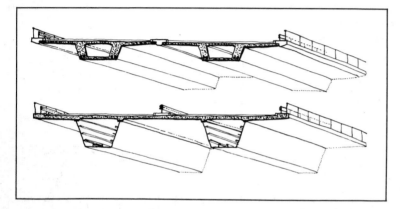

Tubular sections are now commonly used for bridges constructed of steel and concrete. This sketch illustrates typical cross sections for both materials.

and horse traction were used, and the second was the first railway to use steam traction exclusively; it also carried both passengers and goods. In 1824 he accepted an invitation to take charge of railway engineering operations in South America for the Colombian Mining Association of London. A few years later he returned to England and took over the management of his father's factory in Newcastle. In 1845 Stephenson was appointed engineer to the Chester and Holyhead Railway; he therefore was presented with the same problem that had confronted Thomas Telford—the bridging of the Menai Straits. The Admiralty stipulated that a channel for shipping should be kept open during the construction of any proposed bridge, and thus a large span was required.

Since not even temporary staging was permitted in the strait, the adoption of a cast-iron arch constructed on centering was not possible. Further, with the development of the steam locomotive, railway bridges had to accommodate heavier moving loads than road bridges. Although suspension bridges of long spans had been built before the construction of the Britannia Bridge, many engineers considered that a suspension structure could not be made stiff enough to accommodate the increased loadings and vibration.

Several suspension bridges built in the early part of the 19th century collapsed under relatively light loading, and a railway suspension bridge built in 1830 to carry the Stockton and Darlington Railway over the River Tees had a very short life.

Robert Stephenson's solution to the problem was to construct the bridge in the form of a large wrought-iron tube through which the trains could pass. Initially it was thought that the tube would be supported from chains above it, and the masonry piers were built up to the height required for suspension bridge towers.

In order to assess the load-carrying capacity of the tubes, Stephenson consulted another engineer, William Fairbairn (1789–1874). Fairbairn started his career as a mechanical engineering apprentice and in 1811 he moved to London where he was attracted by John Rennie's (1761–1821) Waterloo Bridge, then under con-

General view of the works, September 1848. Work on the piers is almost complete; they were built up to the height required to support the tubes from chains passing through the openings at the top.

A 1:12 scale micro-concrete model of part of the three-lane elevated section of the Mancunian Way.

A model of the 200-foot cantilever arm of the Medway Bridge, a triple cell box section.

struction. However, he was unable to obtain work on the bridge and for the time being remained a mechanic. A few years later he started his own business and soon gained fame as an expert in mechanical equipment.

As part of his work, Fairbairn studied the properties of wrought iron in conjunction with Eaton Hodgkinson (1789–1861). Hodgkinson, who had studied the classic works of French mathematicians, was able to supplement Fairbairn's mechanical expertise with a theoretical background comparable with that of the French engineering schools. A machine for testing materials was developed (Fairbairn's lever) and much research work was carried out with it. The part played by Fairbairn and Hodgkinson in the design of the Britannia Bridge is comparable to the practice adopted by bridge engineers today in retaining the services of specialists to carry out model analysis and testing or site investigations. (Examples of modern bridges in which model testing was involved in the design are the Medway Bridge, the Manchester Skyway Bridge and the Severn Bridge.) After Fairbairn's preliminary tests, it was decided to design the tubes so that they would be strong enough to support their own weight and that of the heaviest trains without assistance from the chains. Bending tests carried out by Fairbairn on tubular sections showed that failure occurred in the compression zone, not the tension zone. These tests were initially made on cast iron, which would normally fail in the tension zone. (Hodgkinson pointed out that the usual bending stress formula could not be applied to thin-wall sections, since thin tubes would usually fail as a result of the compression side becoming wrinkled or buckled before fracture of the tension side occurred. It is only recently that a rigorous analysis of thin-walled tubular sections in bending has become possible, producing a considerable saving in material.) In the light of Fairbairn's preliminary results Hodgkinson suggested that a number of basic tests should be carried out. Because of lack of time, however, Fairbairn was forced to make a decision based on his previous tests in favour of a tube that was rectangular in cross-section. To achieve equal strength in the compression and tension zones of the tube a larger area of material was used in the compression zone. A large model was made, with a span of 75 feet and a rather thin bottom plate. To increase the stability of the compression zone, a cellular structure was used for the upper side. Thus the compression zone was in fact a series of tubes of which the sides interacted to resist buckling, whereas the tension zone consisted of a relatively

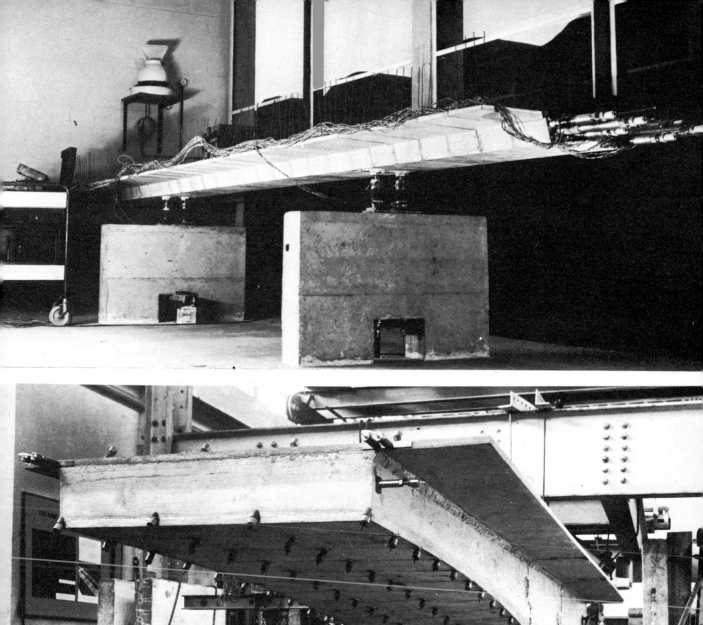

thin plate. In the first test, failure occurred in the tension zone; it was then strengthened in stages until failure occurred simultaneously in the compression and tension zones. At this stage the ratio of the area of compression to that of tension was 12:10. The tests also showed that the vertical walls could fail by wrinkling and additional stiffening was required. (As has been pointed out before, resistance to shear is provided by the vertical walls of the tube.) The final form adopted for the cross-section is shown at right. It is interesting to compare this with a typical stiffened box section used for a modern steel bridge. It is now possible to predict accurately the amount of stiffening required.

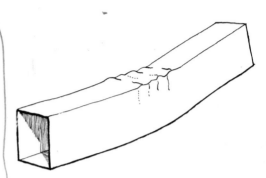

Bending tests carried out by Fairbairn on tubular sections showed that failure would usually occur as a result of the compression face buckling.

An isometric projection of a portion of one of the tubes of the Britannia Bridge. Note the use of vertical stiffeners to prevent buckling of the side plates.

Another interesting feature of the design was the procedure adopted for the erection of the tubular girders. The structure acts as a continuous girder and thus the distribution of moment will be more favourable than for a series of unconnected (simply supported) spans. A comparison of bending moment values for uniform loading on four equal spans is shown on p. 83. However, the end spans of the Britannia Bridge were only half the length of the two centre spans. For uniform loading this situation would produce the moment distribution shown on p. 84. To obtain a more even distribution of moment, since the section remained approximately constant throughout the length of the bridge, the following erection procedure was adopted. In making the connection between sections AB and BC the smaller span AB was tilted before the forming of the riveted joint at B. On lowering span AB to the horizontal position, a bending moment was induced at point B thus giving a more even distribution of moment. A similar procedure was adopted at the other end of the bridge. This procedure can be more easily understood by considering a simpler example. Suppose that it is required to obtain approximately equal moments at points B, C and D in the double cantilever beam shown on p. 84. By applying loads at A and E it is possible to equalise the moments. The magnitude of the load is dependent on the ratio of the cantilever arm to the main span. At the time of construction of the Britannia Bridge this was a very advanced technique. In recent years this technique has been used in the construction of several bridges including Riccardo Morandi's Via Olimpica overpass in Rome.

The site chosen by Robert Stephenson for the Britannia Bridge was in line with the Britannia Rock in the centre of the strait. This ensured a firm foundation for the centre pier. When designing a continuous beam bridge it is essential to minimise the differential

ISOMETRICAL PROJECTION
PORTION OF ONE OF THE TUBES
BRITANNIA BRIDGE.

82

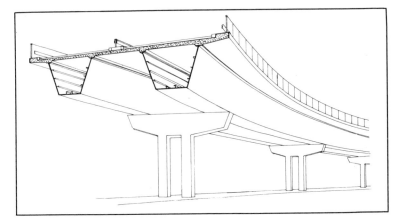

A typical modern steel box section, incorporating stiffeners to prevent buckling of the thin wall plates.

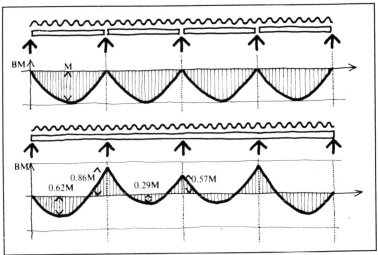

Bending moments can be reduced by making beams continuous. The value of M for the simple spans is reduced to 0·86M for the continuous spans. Note that the maximum bending moment for the continuous spans occurs at the penultimate support.

movement of the supporting piers. If one pier sinks in relation to the others the distribution of moment in the beam will be altered, with the possible result of high overstress. For this reason, continuous girders have generally not been employed for bridge construction except where foundations can be carried down to rock, thus reducing the risk of the piers settling to a minimum.

The foundation stone of the Britannia Bridge was laid in April 1846. Workshops for the fabrication of the tubes were put up on the site together with accommodation for most of the 1500 men employed on the work. Construction began with the foundation work for the

Moment distribution for a four-span continuous-beam system in which the outer span is half the inner span. M represents the simply supported moment for the inner span L.

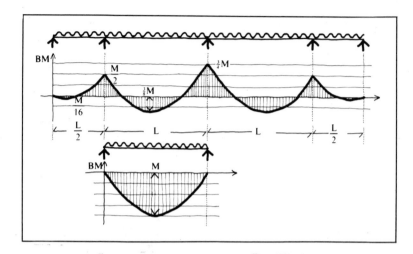

Beam erection procedure for the Britannia Bridge. On lowering the span AB to the horizontal position a more even distribution of moment is achieved.

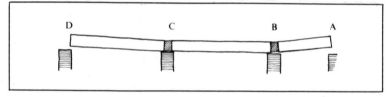

By applying loads at the ends of the cantilever it is possible to equalise the moments at A, B and C.

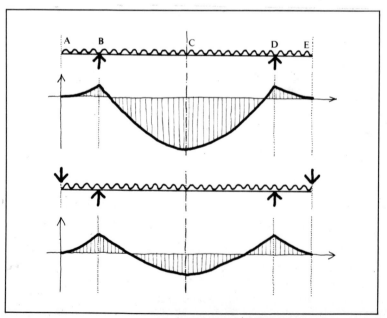

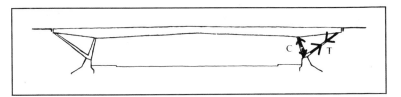

Sketch illustrating the structural principle of the Via Olimpica Overpass in Rome. The vertical component of the tie force T produces a moment distribution to reduce the moment at the centre of the deck, where minimum depth was required.

Construction of the Conway Tubular Bridge, March 1848. The tube is being raised 18 feet by hydraulic lifting tackle.

red sandstone piers, which were to support the two independent tubular girders. At the same time, the fabrication of the riveted tubular girders was started. The shorter 230-foot spans were erected in place on scaffolding. Stephenson was then faced with the major construction problem of erecting the two central 459-foot spans.

Also on the route of the Chester and Holyhead Railway was the Conway River crossing. The Conway Bridge, which was of similar structural form to the Britannia Bridge, was opened in May 1859, a few weeks before the erection of the central Britannia tubes was begun. Stephenson was therefore able to use the construction of the Conway Bridge tubes as a trial run for the much more hazardous

85

task in the Menai Straits. The Conway Bridge has a single span of 412 feet and the tubes had to be raised about 18 feet above water level. Stephenson directed the operations at the site, accompanied by a distinguished onlooker, Isambard Kingdom Brunel. The tubes were fabricated on the shore and then floated out on pontoons to the required position. The tubes, each weighing over 1000 tons, were raised slowly into position by hydraulic lifting tackle. The whole operation at Conway proceeded smoothly, but the Britannia Bridge turned out to be more troublesome. As at the Conway Bridge, the tubes over water were built on the shore and then floated into position on pontoons. Unfortunately a capstan gave way while hauling in one of the guiding ropes and the tubes were in danger of being carried away by the stream. Luckily with the help of some onlookers, this disaster was prevented and the tube was placed on

The 460-foot tubes of the Britannia Bridge, each weighing over 1500 tons, were floated into position on pontoons before being raised into place.

shelves, which had been prepared at the base of the piers. The next stage involved raising the tubes over 100 feet above water level; this was to be carried out using hydraulic lifting tackle. One of Stephenson's young engineers suggested that the lifting operation could be carried out in, at the most, two days.

Stephenson insisted that the tubes should be raised inch by inch and that packing should be placed underneath them at each stage. This was very sound engineering decision, since one of the hydraulic cylinders failed and the tube fell nine inches on to the packing below. The tube suffered only slight damage, and if it had not been for Stephenson's foresight, it could well have been lying at the bottom of the Menai Straits.

All four tubes were erected in a similar manner and connected to the outer spans to form a pair of continuous girders 1511 feet long. The bridge was opened in March 1850 and for five years remained the largest railway span in the world. Robert Stephenson died nine years after the completion of the Britannia Bridge and was buried alongside Thomas Telford in Westminster Abbey.

Today railway passengers travelling to Holyhead probably give the pair of lions at each end more attention than the bridge itself, which may well seem rather ponderous. Yet its design, concept and construction technique earn the bridge a high place on the list of notable engineering achievements during the 19th century. Young engineers are able to learn more by studying this bridge than many others built over a hundred years later.

A pair of lions guards each entrance to the bridge, enhancing the faintly Egyptian air of the whole structure.

87

Squire Whipple's Truss Bridge

The truss is an extremely practical structural system since it can be made from relatively short, lightweight members, which can be built up in sections. Timber trusses were used by the Romans although, not surprisingly, none now survives. This was probably because the practical difficulties of constructing beam-and-arch bridges in stone led the Romans to consider the use of an alternative structural system that did not involve the handling of heavy elements. Indeed, timber bridges using a crude form of truss were probably developed early on in most forested countries. However, because of timber's lack of durability compared with stone, early examples of timber bridges are not easy to trace. The Italian architects of the Renaissance were interested in the use of timber for trussed bridges, and Palladio built several wooden bridges with spans exceeding 100 feet.

The Swiss are traditionally renowned for their carpentry skill, and there are examples of timber bridges in Switzerland dating back to the 14th century. A notable example of 18th-century Swiss timber truss construction is the bridge over the River Rhine at Schaffhausen, completed in 1757. The bridge was built by Hans Ulrick Grubenman (1709–83) and had two spans of 171 and 193 feet. The central stone pier was all that remained of a former structure.

The bridge turned out to be strong enough to carry carriages of up to 25 tons in weight. (Hans's brother Johannes built a bridge at Reichenau of similar form but with a single span of 240 feet.) The design consisted of struts radiating from the abutments, joined by horizontal and vertical members, and the roof was covered with a light timber structure. In 1778 the two brothers built a further bridge at Wettingen over the Limmat near Zurich. This bridge was more closely related to the arch form than the previous two. Unfortunately all three bridges were destroyed by the end of the century.

It is interesting to note that these bridges used the combined truss and arch principle. Although the theory of the simple truss is one of the easiest problems to work out in terms of structural mechanics, a scientific design was not attempted until the 19th century. The analysis of an arch, which is much more difficult, had nevertheless been attempted by several engineers at this time. However, the principle of arch construction had been much more widely used in

A model of the timber bridge over the Rhine at Schaffhausen built by Hans Grubenmann.

Below, a model of the timber bridge at Wettingen. This structure was more closely related to the arch form than the Schaffhausen Bridge.

89

Various types of timber viaduct designed by I. K. Brunel for railways in the west of England, top to bottom: Liskeard Viaduct (reconstructed 1895), Treviddo Viaduct, and Penryn Viaduct.

Interior of the former Shedd's Bridge near Bennington, Vermont, showing the Town lattice construction.

earlier bridges, and the arch was probably incorporated in these early timber structures because of lack of confidence in the truss.

The railway age created a demand for a large number of bridges both in Europe and America. Iron was soon to become the dominant structural material, but timber was still being used in the West of England by Brunel for the construction of viaducts designed for the South Devon, the Cornwall and West Cornwall Railway Companies during the 1850s. Standardised designs were produced with some variations between the three railway companies. The viaducts for the South Devon Company consisted of pairs of trusses carried on masonry piers with spans of about 60 feet. For the Cornwall Railway two main types were developed. 'Fan' viaducts were built across the dry valleys; in these a series of struts radiating fan-wise from masonry piers (in a manner similar to the early Swiss bridges) supported the deck. Where tidal creeks were to be crossed trestle piers were supported on timber piles driven into the mud. Light timber trusses spanned between the trestle piers. The design allowed for any timber member to be removed and replaced in a short time and some of these viaducts survived well into the 20th century. The West Cornwall viaducts used the fan principle but were carried on timber trestles at 50-foot intervals. The use of iron for bridge trusses was not generally successful in the early years of its development as a structural material. Cast iron is notoriously unreliable in tension and this resulted in many failures.

In America the truss-arch principle was widely used at the beginning of the 19th century. In 1820, however, the architect Ithiel Town patented a truss sytem that was completely independent of the arch. The main feature of this truss was an arrangement of closely spaced intersecting diagonals that formed the web. The horizontal members were composed of two or more parallel timbers so that the web

The Howe Truss Bridge, elevation and plan. Initially the vertical members were wrought-iron rods.

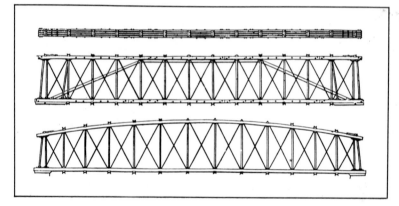

Two typical versions of the Pratt truss, in which the vertical members were timber.

members could be inserted tightly between them. The top and bottom chords were built up of planks approximately the size of those used for modern domestic construction. These trusses were frequently roofed and sometimes covered at the side.

The simple erection procedure, the use of common timber sizes and nailed connections were totally in line with the American philosophy of building. From early pioneer days, American engineers have always appreciated the advantages to be gained from rapid construction methods. This attitude prevails today. Milton Brumer, the engineer responsible for the construction of the 4260-foot Verrazano Narrows bridge stated that during the construction of this bridge 'there was no time for trouble.' Some engineers overlook the fact that speed of construction is an essential feature of bridge

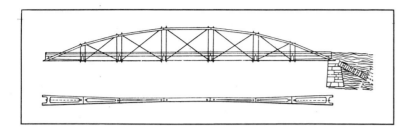

The Whipple arch or bowstring truss.

design, especially so in a modern, industrialised country.

A disadvantage of the Town truss was its tendency to warp because of the flexibility of the web section. Town suggested how his truss would be adapted to cast and wrought iron, but it was not until 1859 that an iron form appeared.

Another pioneer in the development of trusses in America was William Howe. A modified form of the Howe truss dominated timber railway bridge construction for the rest of the century. (Initially, the vertical members were wrought-iron rods, which indicated that Howe had designed them as tension members.) The first scientifically designed truss was the Pratt truss, which was a modification of the Howe form. Thomas Pratt (1812–75) was educated at the Rensselaer Polytechnic Institute in Troy, New York, where he studied architecture, building construction, mathematics and natural science. The Pratt truss had parallel top and bottom chords, which were made of timber. The vertical members were also timber but the double diagonals were of wrought iron. Thus in Pratt's truss the verticals were treated as compression members and the diagonals as tension members. Again, the Pratt truss or a modification of it was to become a feature of American railroads.

The first all-metal trusses to be used in America were designed by Squire Whipple. Whipple was born in 1804 and up to the age of 30 he was involved in several activities including farming and school teaching. He then became interested in railway engineering and in 1841 was granted a patent for what he termed an arch truss. This truss had a polygonal top chord of cast iron and wrought-iron lower chords as diagonals. In 1847 he obtained a further patent for a truss, which was to become commonly known as the Whipple truss. The basic form consisted of parallel chords, inclined end posts, closely spaced intermediate verticals and diagonals, each of which extended across two panels. The posts and top chords were cast iron

94

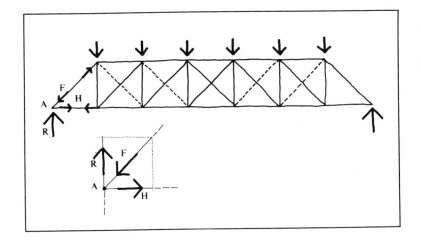

and the other members wrought iron. A modified form was subsequently developed as an all-wrought-iron, pin-connected truss.

In 1847 Whipple published an essay on bridge building, which gave the first scientific analysis of a truss. He appreciated that a body can only be prevented from falling by means of inclined forces. That is, if the weight of the body cannot be counteracted by an equal and opposite force in line with that of gravity, then it can be sustained by a series of inclined forces whose total impact passes through the centre of gravity of the body in a vertically upward direction. It is evident that Whipple understood the concept of force resolution into vertical and horizontal components and the principles of statics. In the analysis of the truss shown above Whipple assumes that the diagonals can only work in compression, and thus it reduces to this simple form shown. It was assumed that the load on the individual members must be axial; in other words, no bending was developed and hence it was possible to consider the equilibrium of the forces acting at each joint. Considering the equilibrium of any one joint, then a force polygon can be assumed to exist around each joint illustrated in the line diagram of the truss. Whipple analysed trusses for the effect on them of a uniformly distributed load, representing the sum of the dead load of the structure, and of a moving load.

Shortly after the publication of Whipple's book, several methods of analysing trusses were developed and European engineers began to study the various truss systems developed in America. Karl Culmann (1821–81), who had been working on the design and

Above opposite, an example of a Whipple bowstring truss bridge (the deck has been removed).

Below opposite, a Whipple parallel-chord truss with inclined end posts. This type of truss was patented by Whipple in 1847.

95

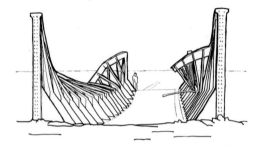

Failure of an open-truss bridge due to lack of stability of the top compression chord.

construction of railway structures for the Bavarian railways, visited England and America in 1849 to broaden his engineering education and published an extensive study of English and American bridges. Culmann was somewhat critical of some of the American truss systems (timber and iron), which he considered to have a small margin of safety. He also demonstrated that the Americans designed their bridges for smaller moving loads than those specified in Europe, and at the same time, used higher working stresses. However, Culmann was of the opinion that the Whipple truss was superior in its rigidity to the other iron bridge trusses he investigated during his American tour. An important feature of the Whipple truss is the use of horizontal bracing in the top and bottom chord planes.

Several failures of open bridge trusses occurred during this period because of lack of lateral rigidity in the top compression chord. The first theoretical investigation of this problem was made by F. S. Jasinsky (1856–99) who designed bridges on the St Petersburg–Warsaw and St Petersburg–Moscow railways.

One of the earliest trusses built in the form shown on p. 97 was the Troy Bridge, New York. Squire Whipple built this bridge for the Rensselaer and Saratoga railroad and it was completed in 1853. The bridge was a single span structure, 146 feet between abutments. The truss depth was 22 feet 9 inches and the verticals were spaced at 10 feet 6 inch centres.

The bridge was built on the skew; as a result, the timber cross-beams between the verticals (posts) were not at right angles to the line of the top and bottom chords. The structure was braced against lateral sway by means of diagonal rods placed in the horizontal plane between the cross-beams. The top chord and posts of the truss were cast-iron, the diagonals wrought-iron rods; the bottom chord was built up of wrought-iron bars. It can be seen from the cross-section through the structure that the top chords are braced in the horizontal plane to achieve lateral stability.

The usual weight of locomotives at this time was 35 tons and Whipple designed the bridge for a moving load of 2000 pounds per linear foot, but in the 30 or so years that followed, locomotive weights doubled, and the bridge had to be replaced in 1883.

An interesting feature of Whipple's book is that he included tables of maximum stresses for timber and iron members. He appreciated the inelastic behaviour of metals at high stresses and listed the stresses to which the metal could be safely exposed in practice. He also recognised the unreliability of cast iron in tension.

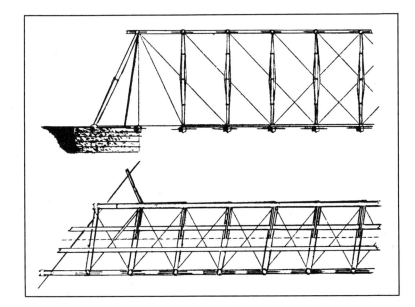

The bridge built by Whipple for the Rensselaer and Saratoga Railroad in 1853 about seven miles north of Troy, New York (elevation and plan).

Since the majority of American bridge builders did not have the benefit of the mathematical training that could be obtained in the European engineering schools, Whipple's clearly reasoned treatise should have been of great value to his contemporaries. Yet it appears that Whipple's work had little immediate influence on practical engineers, although his trusses were to dominate the building of American long-span railroad structures for many years. One of the last Whipple trusses to be built was at Cincinnati in 1899. This was a single-span structure 115 feet long with three parallel trusses, which supported a two-lane road and two main lines. It was replaced in 1937.

Garabit Viaduct

The railway age produced many masterpieces of bridge building; the climax of this great era was reached at the end of the 19th century. In Britain and America very long spans were achieved using the cantilever truss principle, of which notable examples are the Mississippi River Bridge, Memphis, Tennessee completed in 1892; the Forth Bridge completed in 1890; and the Connel Ferry Bridge, Argyllshire completed in 1903.

The development of railways in France lagged behind that of Britain and America; in 1841 there was less than 600 miles of track. However, with the increasing use of iron ore and coal more tracks were laid connecting the mining areas with the main railway routes. Around the Massif Central there are scattered mineral deposits and their development led to the construction of railway tracks in an area with a particularly difficult topography. It was in this area that Alexandre Gustave Eiffel (1832–1923) built, between 1864 and 1884, many bridges that rivalled the achievements of British and American engineers.

Although he was one of the most prolific engineers of the 19th century, Eiffel's achievements in bridge building have not attracted the attention they deserve. A possible reason is that his previous work has been overshadowed by the 1000-foot tower built in Paris in 1889 for the International Exhibition. Baker and Fowler's Forth Bridge and the Eiffel tower are the outstanding monuments of 19th-century engineering skill.

Eiffel attended the Ecole Centrale des Arts et Manufactures in Paris from 1852 to 1855; he then started his professional career as a civil engineer working for various French railway companies. In 1867 he opened the 'Maison G. Eiffel', and in 1880, at the height of his career, he was working on projects in at least a dozen different countries. He was responsible for an enormous range of designs including bridges, dams, locks, gasometers, reservoirs, churches, casinos and department stores. In 1885 he designed the inner structure for the Statue of Liberty in New York Harbour. After being involved in the French Panama canal venture in 1893 Eiffel spent the last years of his life studying aerodynamic problems. He used the Tower for his experiments and built one of the first wind tunnels in 1912. He died in Paris in 1923.

The Garabit Viaduct crossing the Truyère.
Eiffel proposed a parabolic arched viaduct with a rise of about 400 feet.

Two notable examples of railway bridges using the cantilever truss principle: the Mississippi River Bridge at Memphis, Tennessee, and (below) the Connel Ferry Bridge, Argyll.

The Eiffel Tower.

The large viaducts constructed by Eiffel in the Massif Central region of France were mainly in the region bounded in the east by Lyons and in the west by Limoges. Here, between 1864 and 1871, viaducts were constructed at Busseau, Neuvial, Bouble, Rouzat and Bellon. These were of the trussed girder type. Between 1880 and 1884 the Tardes and Garabit viaducts were constructed; the Garabit Viaduct was some distance south of the other viaducts on the line connecting Neussargues with Berziers. But to appreciate the difficulties that confronted Eiffel in the design and construction of these viaducts it is necessary first to consider the character of the railway routes that grew up on the Massif and its unique topography.

The area consists of vast windswept plains extending from narrow gorges. Although the land rarely rises more than 3000 feet above sea level the climate is often severe, as is evident from the architecture of the towns. The railway lines were built to serve the mining districts, which were at the peak of their development. The lines were mainly used by heavy goods trains and the chief geographical obstacle was the number of steep gorges cutting into the uniform plateau of the Massif. Today, the viaducts constructed by Eiffel may look out of proportion to the obviously secondary railway lines. Yet 80 years ago the situation was completely different. Transport from the mining areas in the mountains to the centres of industry and the major railway junctions represented a fundamental problem for the vastly important iron industry.

To build railway lines and viaducts crossing the deep gorges that split the inhospitable Massif Central, Eiffel had to face problems whose solution in an economic manner required the qualities of a master of engineering. Significantly, Eiffel did not study at the Ecole des Beaux Arts or at the Polytechnic. He was therefore not greatly influenced by the theoretical conventions and modes of construction of that period. Eiffel adopted an empirical approach to his work and he never abandoned this attitude throughout his career. He was never in any way influenced by tradition. He always identified the bare essential problem and set about solving it by careful study and analysis of the facts before him. If the solution was not perfect, it meant that he had not examined the problem deeply enough, that he had not grasped the main point, and had not comprehended all the details. The problems and difficulties to be overcome were always the starting points in Eiffel's total commitment to his work; they formed the basis of the challenge, which he then endeavoured to overcome.

101

One of the major problems to be solved in the construction of high viaducts was to produce a structural form that would remain stable under the action of high winds that blew into the gorges. Eiffel carried out a thorough investigation into the effect of wind on structures. It should be remembered that the Garabit Viaduct is as high (480 feet) as the tallest office blocks being constructed in cities today, and it is still not easy to formulate the effects of wind accurately. To help him make a quantitative estimate of the effects of wind on structures, Eiffel set up a network of metereological

One of the Bellon Viaduct piers. In contrast to the Busseau Viaduct the tubular members were widened at the base in parabolic form to increase stability.

Below, Eiffel's first high-piered viaduct, at Busseau sur Creuse. It has six spans; the overall length is about 950 feet and the maximum pier height is 195 feet. Eiffel adopted a reticular structure which sets up the least resistance to the high winds blowing up the Creuse valley.

stations throughout France and over a period of years collected data that might help towards a deeper understanding of wind forces. (In doing so, he rejected all the traditional type of meterological station and devised and constructed new ones.) Eiffel's viaducts and the Forth Bridge were probably the first bridge structures for which a quantitative estimate of wind forces was attempted. The wind pressure on a surface is affected by several factors: its shape, its height above ground level and the degree of exposure. Wind tends to accelerate as it channels along gorges and if the structure is slender, oscillations may develop. Eiffel was aware of both the static and dynamic effects of wind.

Before building the Garabit Viaduct, Eiffel had the chance to develop his bridge engineering skill in the design and construction of several viaducts some way north in the region between Limoges and Gannat. The experience gained in the construction of these viaducts helped Eiffel in the conception of the larger structures at Ponto sur Douro and Garabit. Eiffel's first high-piered viaduct was constructed at Busseau in the Creuse Valley. This was a six-span structure with an overall length of about 950 feet, the maximum height of the piers being 195 feet. The high winds channelling through the Creuse valley would impose large pressures on a solid obstacle; for this reason, Eiffel chose a reticular truss for the main beams, which would set up least resistance to the wind. The piers were also of open form with the width at the base facing the wind's principal direction along the valley much greater than at the top. Thus the form of the structure was directly related to the effects of wind, with the width of the structure increasing as the overturning effect of the wind increased. The main trusses support a two-way railway track on which the speed of the trains is limited to about 20 miles per hour. The line was opened in 1864.

Between 1867 and 1869 Eiffel adopted a parabolic pier for the viaducts on the Gannat-Commentry line. An example of this is the pier for the Bellon Viaduct. In contrast to the Busseau Viaduct piers, the tubular members were widened at the base in parabolic form. If such a pier is thought of as a vertical cantilever, then its form near the base is similar to that shown in the diagram of bending moment induced by uniform loading along its length. Although an assumption of uniform loading produced by wind is a gross oversimplification, the logic of this form of construction can be appreciated.

This idea was later used in the design of the Tower. (For the Garabit Viaduct piers, Eiffel used a truncated pyramid-type of construc- 103

tion similar to the Busseau Viaduct. The reason for this was the gradual changeover from the use of cast and wrought iron to steel.) The ease with which cast iron could be formed into curved members made the use of a parabolic base economically feasible. The economic use of steel depended on the production of straight members of standard section. However, the manner in which Eiffel detailed his various wrought- and cast-iron members anticipated, in both profile and general shape, the up-to-date design of steel members. In the mid-1870s the price of steel on the world market was very high, but in the following ten years it fell by 75 per cent.

The materials used for the construction of the high piers for viaducts on the Gannat-Commentry line were cast and wrought iron. Since double diagonal wind bracing was adopted, they could be designed as tension members, but the difficulty which Eiffel had to overcome was the connection between the diagonal and vertical members. Traditionally the problem was resolved by binding cast-iron elements with large loops and clips to which other members could be attached. The manner in which Eiffel detailed his connections is shown on p. 106, which is a view looking up one of the piers

Left, the powerful crescent-shaped arch of Eiffel's viaduct carrying the railway across the Truyère at Garabit.

The base of one of the Bouble Viaduct piers, with curved outer tubes to increase the stability of the structure.

105

of the Bouble Viaduct. The pier consists of four corner pipes in cast iron braced with U-shaped diagonals in wrought iron. A spiral stair is supported by a fifth cast-iron pipe in the centre of the pier. A close-up of the connection between the vertical tubes and the diagonal members is shown at right. The vertical tubes are lipped or flanged at the ends so that a bolted connection can be made between the adjacent sections.

The U-shaped diagonal wind-bracing members are connected to the tubes by rivets attaching them to the gusset plates. This technique anticipates modern steel detailing procedures although the introduction of welding has enabled much neater connections to be adopted.

The section at the foot of the Bouble Viaduct piers is characteristic of the Gannat-Commentry line; the curved outer pipes are similar in form to the base of the Eiffel Tower.

Ten years after the completion of the Gannat-Commentry line, a railway link between Marvejols and Neussargues was planned. This was to connect the southern areas of the Massif Central with the southern line and the Paris-Lyons-Marseilles line. There were many difficulties to be overcome along this route, which was never less than 2000 feet above sea level and for long distances over 3000 feet. Its problematical character led the official engineers who were studying the project to consult Eiffel over a provisional design. Eiffel suggested that the Truyère should be crossed in the Garabit locality, where the height of the viaduct would be about 400 feet above water level. Eiffel proposed that the Truyère should be spanned by a parabolic arched viaduct of the type completed a year previously over the Duoro at Porto. It was estimated that the construction of this large viaduct would result in a saving of about 50 per cent produced by the shortening of the track and the elimination of other viaducts. In June 1879, the Ministry gave Eiffel direct responsibility for the work.

The Garabit Viaduct was completed in 1884 but the other sections of the line were not finished until 1888. In the late 1870s the price of steel had dropped considerably and thus this material was adopted for both the Douro and Garabit viaducts.

Because of the great height of the Garabit Viaduct, the effect of wind loading was even more crucial than it was in the Gannat-Commentry viaducts. Eiffel had to take into account a wind force of 55 lb per square foot for stationary trains and 30 lb per square foot for moving trains. On this line, trains were stopped in the

106

One of the Bouble Viaduct piers. The double diagonal bracing members adopted for wind bracing were also designed as tension members.

The arch was constructed in the form of a crescent-shaped moon (right).

The connection between the vertical and diagonal members. The vertical tubes are lipped at the ends so that a bolted connection can be made. The diagonal members are attached to gusset plates – anticipating modern steel detailing techniques.

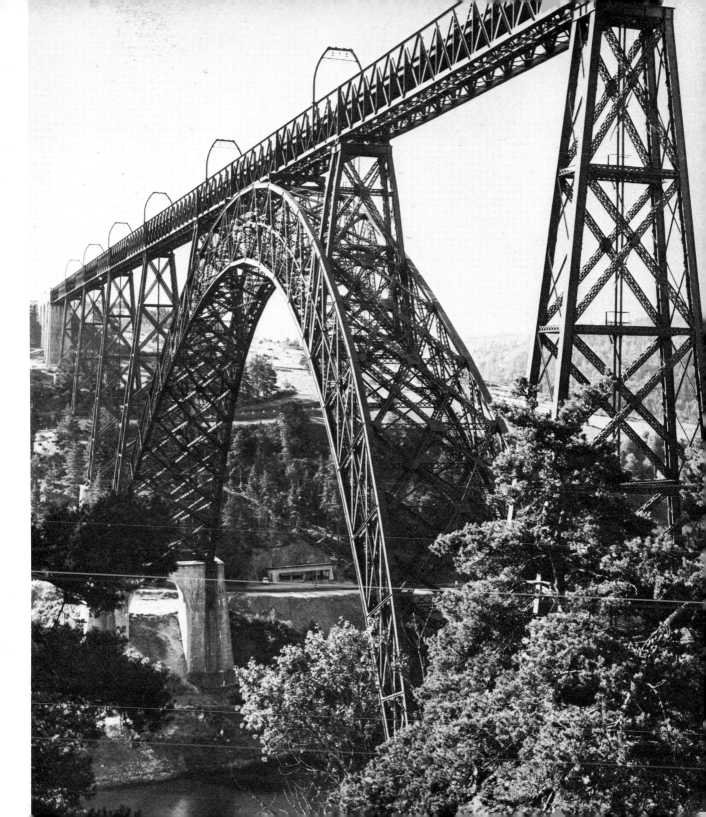

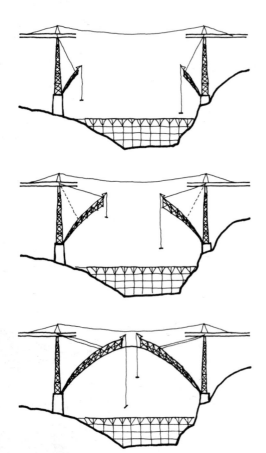

The arch ribs were constructed by the cantilever method: the ribs were built out in sections and tied back to the approach viaduct.

stations and the service suspended if the wind speed produced a pressure greater than 30 lb per square foot.

The form adopted for the parabolic arch was based on the same reasoning that produced the extended parabolic-based piers for the Gannat-Commentry line. The arch, span 530 feet, was supported on hinges at the base and was in the form of a 'crescent moon' in elevation. The progressive increase in thickness from the hinges to the crown was made necessary by the moments induced by asymmetrical loading, the maximum thickness coinciding with the regions of maximum moment. The two arcs that form the truss were placed in oblique and opposing planes, giving a width of 21 feet at the top and 65 feet at the base. This increase in the width of the arch in plan from the crown to the base was necessary to achieve stability against the high winds blowing along the valley.

The width of the base of the piers that have a common seating with the arch was dictated by the width of the base of the arch— that is, 65 feet. The piers were of the truncated pyramid type, 16 feet wide at the top.

The erection procedure for the arch adopted by Eiffel—to cantilever out from the arch-springing in sections, which were tied back to the approach viaduct—is now a common technique for both steel and concrete bridges. The two halves of the arch built out from the hinge positions were connected in April 1884.

When the line was completed four years later, the bridge was inspected. It was found that the deflection of the arch was about one third of an inch under a trainload of 400 tons. In contrast to the smaller viaducts on the Gannat-Commentry line, the rail level was kept some distance below the level of the top members of the truss supporting the track. This served the double purpose of reducing the area exposed to the wind and reducing the risk of a catastrophe brought about by derailment, which had occurred on the Rouzat Viaduct.

The arch increases in width from the crown to the abutments to achieve stability against wind.

Below, the rail level is some distance below the level of the top members of the truss supporting the track. This reduces the area exposed to wind and avoids the risk of catastrophe if a derailment should occur.

Forth Bridge

The majestic Forth Bridge (completed in 1890), spanning the Firth of Forth at Queensferry, symbolises the tremendous achievements of Victorian engineers and the immense strides made in the technique of bridge design and construction since the dawn of the Railway Age some eighty years before. The construction of this bridge coincided with the almost total eclipse of cast and wrought iron. Except for a few hundred tons of cast-iron washers, anchor plates in the piers, and about 2000 tons of kentledge (ballast) in the form of cast-iron bricks laid in asphalt, the whole of the superstructure between the granite piers at each end was built of mild steel. (Cast iron, although easier to pour into shapes than wrought iron, is a brittle material and difficult to produce in uniform quality. It is notoriously unreliable in tension and it would be dangerous to use in bridge construction at a stress greater than a few thousand pounds per square inch. Wrought iron is a much purer and more malleable material than cast iron but is expensive to produce.)

The steel used for the Forth Bridge was manufactured by the Siemens Martin open-hearth process and had the advantage of uniform quality and precisely calculable properties. Extensive tests were carried out before and during the construction of the bridge to establish the reliability of the material and the following rules were laid down on the admissible stresses. These took into account fatigue—that is, the lowering of the breaking point of a member when it is subjected to the repeated changes in load common in railway bridges. For members in tension the steel was to have a breaking load intensity of between 30 and 33 tons per square inch. For members in compression a breaking load of between 34 and 37 tons per square inch was required. Where the load frequently varied between nil and maximum, then the breaking load was to be taken as 20 tons per square inch, increasing to $22\frac{1}{2}$ tons per square inch if the variation occurred only rarely.

For loads that produced alternately tensile and compressive effects, the breaking load was taken as 10 and 15 tons per square inch for frequent and occasional variations respectively. The working load was to be one third of the breaking load. Rules were also laid down concerning the elongation of test specimens under load to ensure adequate ductility.

The Forth Bridge serves several important railway connections. Before the bridge was built, the Firth was crossed by three steam ferries, Granton to Burntisland, South to North Queensferry and at Kincardine. The first railway bridge was at Alloa 20 miles from Queensferry. Passage by ferry was frequently affected by weather conditions, especially the most seaward crossing, while travelling by rail via Alloa or Stirling added well over one hundred miles to the journey. The construction of the Forth Bridge therefore considerably reduced rail travel time from Edinburgh to Perth, Dundee or Aberdeen.

The Forth Railway Bridge.

111

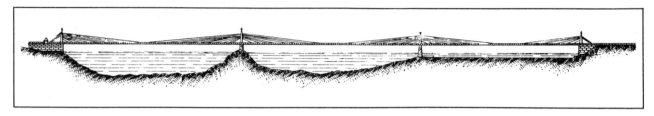

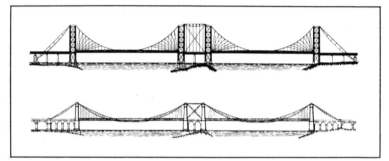

James Anderson's design for a bridge over the Forth, 1818.

Left, alternative preliminary designs for the Forth Bridge by Sir Thomas Bouch.

The first documented proposal for spanning the River Forth was prepared by an Edinburgh civil engineer, James Anderson, in 1818. His report was accompanied by diagrams illustrating some rather slender chain bridges with a main span of up to 2000 feet. About this time work was begun on Telford's Menai Bridge (see p. 67). As can be seen above Anderson's design was very primitive compared with Telford's. It was in fact abandoned and if it had been built, it is likely that the structure would have soon disappeared into the sea. In 1860 the North British Railway Company fixed a site for a bridge consisting of a number of 500-foot spans in the centre of the structure with approach spans of shorter length. This project too was abandoned, and in 1873 the Forth Bridge Company was formed for the purpose of constructing a suspension bridge with two spans of 1600 feet. The structure was designed by Sir Thomas Bouch, who had also designed the trusses for the 1860 proposal and other bridges including the ill-fated Tay Bridge. The Tay Bridge collapsed in December 1879. Bouch was probably given an unfair share of the blame for this disaster, which was mainly due to bad workmanship. Nevertheless, confidence in Bouch's suspension-bridge design was shaken and work on the structure was abandoned. The Forth Bridge Company then appointed Barlow, Harrison, Fowler and Baker as consulting engineers, who carried out a comprehensive feasibility study for various structural systems. It was finally decided to adopt a cantilever design, which was a relatively new structural system for railway bridges although the antiquity of cantilever construction rivals that of the arch and chain. And some 20 years before the construction of the Forth Bridge an American engineer, C. H. Parker, was designing iron-truss cantilever bridges of the type shown at right. The trusses were anchored down by a series of tension rods.

A sketch illustrating cantilever trusses designed by the American engineer, C. H. Parker. The trusses were anchored down by means of tension rods.

The earliest example of a cantilevered structure is probably the stone corbel-and-lintel combination in Egyptian and Indian temples. Here the wooden deck structure is corbelled out from masonry piers. A development of this idea was the use in India of large timbers laid longitudinally one above the other, each timber being somewhat longer than the one below. The advantage of this form is that bridges can be built over deep gorges and large spans with the minimum of intermediate support. Although much greater spans could be achieved with suspension bridges, 19th-century engineers found it difficult to solve the problems of stability associated with this form of construction, especially for railroad requirements. The principle of the cantilever construction for the Forth Bridge was clearly demonstrated by Mr Benjamin Baker at a lecture, using the human body to simulate the manner in which the principal stresses are distributed. Two men sit on chairs with their arms outstretched, grasping sticks, which are butted against the chairs. The men therefore represent two complete piers, illustrated in the outline drawing above their heads. The central girder is represented by a stick suspended or slung from the two inner hands of the men, while the anchorage provided by the counterpoise in the cantilever-end piers is here represented by a pile of bricks at each end. When a load is put on the central girder by someone sitting on it, the men's arms and the anchorage ropes come into tension while the men's bodies from the shoulders downwards and the sticks come into compression. The chairs represent the circular granite piers. Imagine the chairs one third of a mile apart and the men's heads as high as the cross of St Paul's, their arms represented by huge lattice steel girders and the sticks by tubes 12 feet in diameter at the base. A very good idea of the structure is then obtained.

The Forth Bridge, with a main span of 1700 feet, is still the second largest of its type in the world. It was overtaken in 1917 by the Quebec cantilever bridge with a main span of 1800 feet.

The contract for the construction of the bridge was signed in December 1882. Extensive preparatory work was then carried out;

this included the erection of offices, workshops and stores and the setting out of a base line so that the positions of the masonry piers could be fixed. The bridge consists of two approach spans and the main cantilever structure. Starting at the south end there are four masonry arches, which terminate in the abutment for the south approach viaduct. This is followed by ten girder spans leading up to the south cantilever end pier. The main bridge structure consists of three double cantilevers and two central connecting girders. The north cantilever end pier is followed by five girder spans and then three masonry arches.

The entire structure in elevation is shown on pp. 118–9; the length is 5349 feet 6 inches between the centre lines of the end cantilever piers and the rail level is 157 feet above high water. This gives a clear headway for shipping below a height of 151 feet. (This would be reduced by $3\frac{1}{2}$ inches by the load of two trains.) The balance of the structure is achieved in the following way. The Fife and Queensferry piers are identical and each of the cantilever arms project 680 feet. The addition of the central girder of 350-foot span produces an out-of-balance load equal to half its own weight. It is therefore necessary to load the ends of the outer cantilevers with the equivalent of half the weight of the central girder. This is achieved by making the endposts of the cantilevers in the form of a large box filled with the appropriate amount of dead weight. The balance is then upset by the passage of a train over the bridge. This is counteracted by adding further dead weight to the boxes—enough to counter-balance any possible train load with some to spare.

The central (or Inchgarvie) pier carries half the weight of each of the central girders; it is therefore exactly balanced for dead load. This balance would be upset when a train passes, or worse still, when two trains meet at one end of the central girder. This would tend to make the structure rotate about pier A and to lift the structure off pier B. The structure is fixed to the piers by holding-down bolts and, since it was intended that these should never be brought into tension, the distance between piers A and B had to be such that the holding-down bolts were always in compression under the most unfavourable loading conditions.

Investigations carried out by Baker led to the decision that the principal compression members, the vertical columns of 12-foot diameter, the bottom members of the cantilevers and so on, should be tubular in form. The tubular section is a most efficient compression member since it is light in weight and has a high resistance to

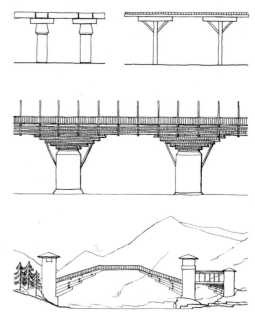

Early examples of cantilever construction in Egypt (top), India, and China.

buckling. Today tubular members are widely used for trusses and also for large beams, where buckling of the compression zone can also occur in bending.

In a structure of this size great care must be taken to allow for changes in length brought about by variation in temperature. If provision is not made for the structure to expand or contract, severe overstress will occur. The fixed points of the structure are located as follows:

1. the south east circular pier of Fife
2. the north east on Inchgarvie
3. the north east on Queensferry.

The expansion lengths from the fixed points are indicated on p. 118.

Allowance also had to be made for wind pressure. The Tay Bridge disaster in 1879, shortly before work started on the Forth Bridge, led Fowler and Baker to carry out an extensive series of tests to determine the effect of wind on exposed surfaces. As a result, the structure was designed for a wind pressure of 56 pounds per square foot.

Work on the bridge itself began with the construction of the foundations for the Fife and Queensferry approach viaduct piers and the Fife, Queensferry and Inchgarvie circular piers, which support the cantilever structure. This work had to be carried out in water and so a temporary dam—a cofferdam or caisson in civil-engineering terminology—had to be used. By using a cofferdam,

A living model illustrating the principle of the Forth Bridge demonstrated by Mr Benjamin Baker. This gives a very good indication of the equilibrium of the structure.

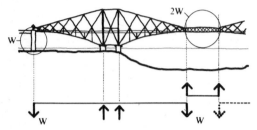

Sketch illustrating how the balance of the Fife and Queensferry piers is achieved.

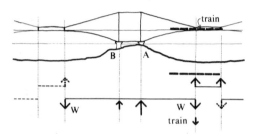

The central (Inchgarvie) pier is balanced for dead load. The distance between piers A and B is such that the structure is stable under the heaviest train loads.

The Forth Railway Bridge is a masterpiece of Victorian engineering, so massive it dwarfs the trains that run across it.

which encloses the whole working area with a perimeter structure from which the water is later removed, foundation construction can be carried out in dry conditions. This device has been used since Roman times, when the sides of the cofferdam were formed by driving oak stakes into the ground below the water. This method is described by Vitruvius in his ten books on architecture. However, there are certain conditions in which a cofferdam cannot be used. A rock stratum or a great depth of water makes the use of piles impractical and a pneumatic caisson must be used. A caisson has a working chamber at the bottom from which water is removed by compressed air. The use of caissons for constructing bridge foundations in the 19th century led to considerable loss in human life. The men worked in a high-pressure atmosphere and if returned too quickly to normal pressure suffered severe pain and even paralysis (known as 'the bends'), fatally in some cases. Other hazards were 'blow outs' caused by compressed air in the caisson escaping through a weak spot in the ground, and fires caused by oil lights in working areas. The introduction of electric light in the construction of the Forth Bridge vastly improved the working conditions. Caissons were used to build the foundations in deep water—the structure being built on shore and floated out into position. The caissons were then filled with concrete to the low water level and at this stage the granite courses were commenced. The piers for the main towers were built up to a height of about 18 feet above high-water level and at this stage the erection of the superstructure could begin. The load from the towers was transmitted to the piers by means of bedplates. The bedplates, consisting of several layers of steel plate, were securely anchored to the piers by holding-down bolts. Work on the vertical columns was now begun, and at a height of about 50 feet above bedplate level a lifting platform was constructed to enable the towers to be built to their full height. At this stage, a start was made on the erection of the bottom members of the cantilevers. As soon as the vertical columns had reached their full height, work on the top members of the cantilevers was started. In the meantime, the approach viaduct piers were being raised. The girders were erected at a convenient level and then lifted at the same time as the building of the masonry piers.

The final stage in the construction was the connection of the central girders and the cantilevers. Referring to the expansion diagram it can be seen that to allow for longitudinal expansion, the central girders were fixed at the Queensferry and Fife ends. Since the

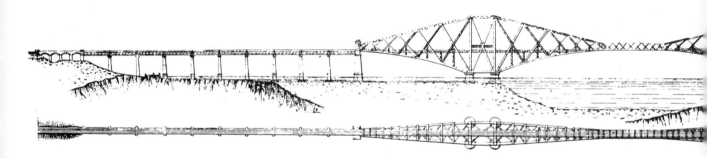

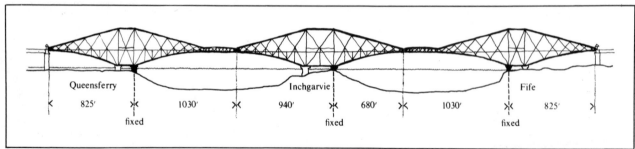

Queensferry   Inchgarvie   Fife

825'		1030'		940'		680'		1030'		825'
	fixed				fixed				fixed	

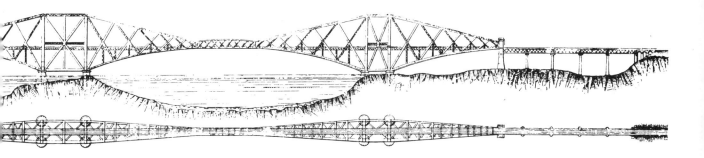

connection of the central girders was to be made at midspan, both ends were temporarily fixed to the cantilevers until they were joined at the centre of the span. The expansion ends could then be released. The first stage was to erect temporary girders, which were attached to the bottom members of the cantilevers. A half bay of the central girder was built on the temporary girders, then moved back and connected to the cantilever.

Work on the girders was then continued using cranes which had moved downhill from the cantilever arms. The connection of the two halves of the central girders depended on the surrounding air temperature. The lengths of the bottom booms were fixed so as to leave a gap of 4 inches between the ends at a temperature of 60 degrees. On October 10, 1889, when the temperature was about 55 degrees, alignment was achieved with the west boom but the gap on the east boom was still about a quarter of an inch, even with the help of hydraulic rams. A fire was therefore lighted under the bottom booms on the east side and the boom then expanded the required amount. The bolts were inserted into the holes and screwed down hand-tight. The structure was test loaded with two 900-ton trains on January 21, 1890, and formally opened by the Prince of Wales on March 4, 1890. Knighthoods were conferred on John Fowler, Benjamin Baker and William Arrol.

The Forth Railway Bridge is still admired by engineers throughout the world, and Riccardo Morandi, the designer of the Maracaibo Lake Bridge and many other notable structures, considers it to be one of the finest and most important works of our time. In 1965 the Queen opened the Forth Road Bridge, a suspension structure with a main span of 3300 feet, about one mile west of the railway bridge. These two great bridges symbolise the achievements of 19th- and 20th-century engineers, one representing the climax of the railway age and the other the age of the motor car.

The entire structure of the Forth Railway Bridge.

Centre opposite, floating out a caisson for the Queensferry pier, 1884.

Below opposite, an illustration showing the expansion lengths from the fixed points.

A pneumatic caisson.

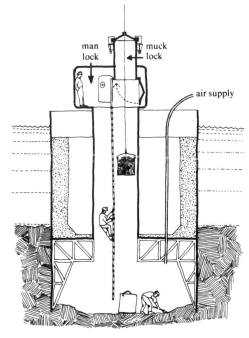

119

Salginatobel Bridge

Today concrete is the dominant structural material; its use as a major material for bridge construction dates back to the latter part of the 19th century. Between 1887 and 1891, for instance, the German firm of contractors Wayss & Freitag constructed a large number of concrete arch bridges with spans of up to 130 feet. Another early example of concrete bridge construction is the Glenfinnan viaduct, completed in 1898, whose structure consisted of 21 plain concrete arches of 50-foot span. In anticipation of a differential movement of the piers, a sliding joint was incorporated at the crown of each arch. Differential movement could cause extensive cracking in the arches since there was no reinforcement to resist the tensile stresses. And at the beginning of the 20th century, Robert Maillart (1872–1940) completed the first of a series of reinforced-concrete arch bridges, which were far in advance of their time. One of these was the Salginatobel Bridge near Schiers in Switzerland, completed in 1930. But before describing it in detail, it is necessary, in order to appreciate fully the quality of Maillart's work, to outline briefly the development of reinforced concrete up to the beginning of the 20th century.

Concrete is usually made by mixing cement with sand, crushed rock and water. The cement combines chemically with the water to form a cement paste matrix around the sand and crushed rock (known as aggregate). The cement paste gradually hardens, forming a material that is strong in compression and weak in tension. Although a cement-like mixture of lime paste and volcanic ash was used by the Romans, modern cements were developed from the experiments of Joseph Aspdin, who in 1824 patented a new variety, which he produced by burning limestone and clay together in his kitchen stove. Concrete is now used in greater quantities than any other structural material and as a result the manufacture of cement is one of the world's largest industries. Even so, the extremely complex reaction between cement and water is still not fully understood. In addition, the gradual hardening of the cement paste involves volume changes in the concrete, known as shrinkage. The amount of shrinkage which concrete undergoes depends on so many variables that it is still not possible to predict its extent with great accuracy. As the hardening process lasts for a long period of time, the concrete is under load before it is fully dried out so that extra

The Salginatobel Bridge demonstrates how Robert Maillart freed himself from the traditional forms of building.

121

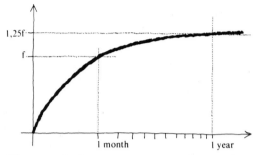

The strength of concrete increases with time; the rate of hardening decreases rapidly after a period of about a month (f = strength in compression).

Opposite, an early example of concrete bridge construction: Glenfinnan Viaduct, Scotland, completed in 1898 (see also p. 41).

water tends to be squeezed out of the pores of the material. This phenomenon, known as creep, results in a gradual deformation of the concrete.

The wet concrete is poured into forms, which are constructed to the profile required and supported by staging. When the concrete has reached the necessary strength to accept the load, the formwork is stripped and the staging removed. When designing a concrete structure, careful consideration must be given to creep and shrinkage and the problems involved in the erection and removal of the forms. The quality of Maillart's work is largely the product of his understanding of the nature of concrete as a structural material and of the problems involved in placing it in position.

One of the first users of concrete on a large scale was François Coignet during the 1850s. Coignet realised that excessive water in the concrete mix was a source of weakness and that the material must be adequately compacted in the forms. On the other hand, although he used iron rods and joists in his concrete work, he did not appreciate the composite action of the two materials, the concrete resisting compression and the steel resisting tension. However, after a series of experiments conducted in 1871–2, an American, W. E. Ward, came to the conclusion that iron should be placed near the bottom of a beam to resist the tension. Another American, T. Hyatt, while he was working in England, published in 1877 an account of some tests carried out on reinforced slabs. The slabs were reinforced by $2\frac{1}{2}$-inch by $\frac{1}{4}$-inch iron ties, 6 inches apart and placed at the bottom. Hyatt demonstrated that the compressive strength of the concrete above the neutral axis (see p. 124) was more than enough to counteract the tensile stress of the bars below. The engineers Lambot and Monier have been popularly credited with the invention of reinforced concrete, but there is little evidence to suggest that they understood the true function of reinforcement. By 1886 German engineers had developed theories of beam and slab design and data were published based on their findings.

Between 1879 and 1891 the French engineer François Hennebique (1842–1921) carried out extensive research on slabs, beams and columns. He introduced stirrups as a bonding medium between the compression and tension zone of concrete beams and slabs and bent up reinforcement near the supports. He also substituted steel for iron reinforcement. Hennebique took out patents for his inventions in France and Belgium in 1892 and established himself as a consulting engineer. Concessions were granted to contractors to

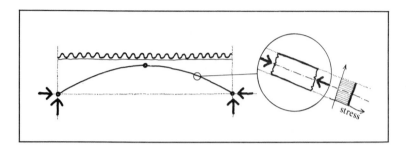

*A parabolic arch subjected to uniform loading—
no bending stress in the arch rib, uniform
compression.*

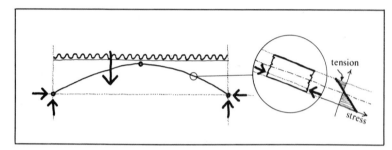

*Non-uniform loading produces bending which
may result in tension developing. Reinforcement
is required to resist the tension.*

operate his patents and Hennebique quickly built up a highly efficient
organisation. Several textbooks on the principles of reinforced-
concrete design were published by members of his staff and an
example of Hennebique's work in the civil engineering field is the
bridge over the Ourthe at Liège in Belgium, completed in 1905. The
bridge has a span of 180 feet, and was one of the few structures built
at this time that could compare with Maillart's work both technically
and aesthetically.

In 1872 at Berne. He was educated at
Zürich Polytechnic between 1890 and 1894, and received a diploma
of structural engineering. Maillart started his professional career in
railway construction, and in 1897 was appointed assistant engineer
to the Highways Department of the Zürich City Council. At the
time of his appointment, designs in steel were being prepared for the
Stauffacher Bridge in Zürich; this was a road bridge over the River
Sihl. Maillart produced an alternative design in reinforced concrete
(at an estimated cost of two thirds of the original), which he was
allowed to carry out and complete. This bridge underwent considerable ornamentation, for which Maillart was not responsible.

In 1901 Maillart set up his own firm of contractors; one of his first
undertakings was a sanatorium for which François Hennebique was
the consultant for the concrete work. There is little doubt that

Maillart gained much experience from his association with Hennebique, but he soon departed from the post-and-lintel basis of early
concrete work, as is shown by his design of the Inn Bridge at Zuoz
(1901). This bridge was one of the prototypes for a series of bridges
to be built over a period of 40 years.

A development from the Inn Bridge was the Tavanasa Bridge,
in which an open spandrel was adopted. Maillart also produced
a new structural form for buildings—the mushroom slab—and
this led to contracts in several countries including Russia, where
he transferred the headquarters of his practice. He returned to
Switzerland in 1919 and set up practice as a consulting engineer.
Maillart, like Freyssinet, was not impressed with the then-accepted
methods for calculating reinforced concrete, which were generally
based on the assumption that Hooke's Law was obeyed. (Hooke's
Law states that in an elastic material, load is proportional to
deformation.) Both Maillart and Freyssinet carried out tests demon-
strating the non-Hookean behaviour of concrete because of the creep
and shrinkage effects described earlier. Maillart preferred simple
structural systems for his bridges, in which the effects of creep,
shrinkage, changes in air temperature and differential movements of
the foundations were reduced to the minimum. As a result, Maillart's
work was criticised by other engineers, who adopted the laborious,

125

Robert Maillart's bridge over the Inn at Zuoz, completed in 1901. The forms in which the concrete is poured are supported on staging.

so-called 'exact' calculations based on Hooke's Law while completely ignoring the actual behaviour of reinforced concrete under load. Maillart also experienced considerable difficulty with municipal and cantonal authorities, so that most of his work was relegated to remote areas. Many of his bridges were constructed in hidden Alpine valleys, which were difficult to get to and which posed formidable problems of construction. If it had not been for the economy with which he could build his bridges, using the materials in the most efficient manner, his designs would probably have been ignored.

The Salginatobel Bridge, spanning a precipitous Swiss gorge,

The Stauffacher Bridge over the River Sihl in Zurich, designed by Maillart.

was constructed between 1929 and 1930. The mountainous location posed a difficult scaffolding problem requiring great structural ingenuity for its solution. The principal components of the structure are the arch of 292-foot span and the deck of an overall length of 432 feet. The arch is hinged at the crown and at the abutments, so that it can be analysed by using simple statics. Another result was that the structure would not be significantly affected by changes in geometry caused by creep, shrinkage and so on. The concept of the arch as an inverted chain has been outlined in the introduction. However, in the Salginatobel Bridge the loading cannot be con-

127

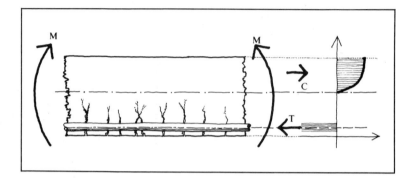

The mechanics of a concrete section subjected to bending (M). The concrete resists the compression (C) and the steel the tension (T). The reinforcement also controls the crack widths.

Opposite, Maillart's bridge over the Salgina gorge posed difficult construction problems. The arch span is 292 feet and the overall length of the structure is 432 feet.

sidered to be distributed evenly over the entire span. Maillart's bridges were deliberately designed to use the minimum amount of material and were never of very long span; thus the live loading represented a significant proportion of the total load. This meant that bending stresses would be induced and the arch section was designed to resist these stresses. It was therefore necessary to consider the live loading in several positions to establish the maximum bending moment at each section of the deck. Where tension occurred, reinforcement was introduced to resist the tension. Since the resistance of concrete to tensile stress is low (a few hundred pounds per square inch) the concrete will crack whether or not the reinforcement is provided; reinforcement will, however, limit the width of cracks. Shrinkage and temperature differentials over the depth of a concrete section may also produce cracks and therefore create a further need for reinforcement. The reinforcement must be provided with adequate cover to prevent corrosion. The width of cracks in concrete is largely determined by the stress in the steel and the amount of cover. If the crack width is more than about 0·2 mm, moisture will seep through to the steel. Maillart adopted a uniform pattern of reinforcement which controlled cracking in any part of the structure.

The cross-section adopted for the arch corresponds to the variation in bending moment along the arch. Because the arch and the roadway fuse towards the centre, a shallow box section was constructed at this point. This is a characteristic conception of Maillart's; the deck forms part of the structure, so that the amount of material required is cut to the minimum. Continuation of the box section through to the abutments would involve a considerable wastage of material. The deck and the arch are therefore separated roughly at

quarter span so that the box section loses its top flange. As the bending moment reduces towards the abutments, so the double T-section grows smaller. Between quarter span and the abutments, the load from the deck is transmitted to the arch by means of three vertical walls.

The deck is a continuous slab lying over the walls, and the pattern of reinforcement is similar to that developed by Hennebique. The use of fillets at the junction of the slab and the walls provides a local thickening that corresponds to the increased bending moments occurring over the intermediate supports of the continuous slab-and-beam systems. The economic reasoning behind this idea is somewhat dubious, because of the additional formwork costs incurred in providing the fillets. However, a further justification for the use of fillets is that local stress concentrations are avoided and a smooth stress flow between the various parts of the structure is ensured. This is of course a sound engineering principle, but its realisation is difficult in a material such as concrete, which has to be poured into moulds.

The positions in the structure at which stress concentration occurs are at the three hinges. To ensure that rotation can take place, the section of the hinges is made as narrow as possible. The stresses are therefore channelled through a small area, and reach a relatively high level. The reinforcement is scissored at the hinges and mats are placed each side of the hinges to prevent bursting.

The Salginatobel Bridge demonstrates how Maillart freed himself from the traditional forms of building in a way that allowed him to use concrete in a completely rational manner. (It is interesting to speculate whether, while doing so, Maillart was influenced by the Great Stone Bridge in China constructed over 1400 years before.) Not until three years after the completion of the Salginatobel Bridge was reinforced concrete used in Britain for the construction of a major railway bridge. Most reinforced concrete bridges built in Britain and America up to 1940 were ponderous in design and often suffered from severe cracking, proving Maillart's belief that mass was no substitute for quality. Some of these bridges were even made to look like castles. Concrete has produced some of the world's ugliest bridges, whose only merit is that they encourage keener appreciation of Maillart's elegant and rational structures hidden in the mountain ranges of Switzerland.

Many reinforced bridges are ponderous in design, and ornamentation in no way enhances their appearance (bridges over the Merritt Parkway, Connecticut).

Esbly Bridge

Before the development of prestressing techniques large-span concrete bridges were invariably of arch form. But although the arch form is suited to materials of relatively low tensile strength, it has several disadvantages. For example, it needs expensive centering and may also involve a considerable wastage of material, since shipping regulations require lanes to be of constant height. If a large opening is to be spanned by a simple beam then the necessary construction depth may be too great. Further, a long-span reinforced-concrete beam requires a great deal of reinforcement. The simple beam solution can be improved by enclosing the opening with a rectangular frame (portal frame), which results in a more favourable distribution of moment and therefore a reduced structural depth. It is, however, more complicated to analyse and build than a simple beam. This form of structure was used for the Esbly Bridge designed by Eugène Freyssinet (1879–1962)—a shallow portal frame bridge over the River Marne in France.

The principle of a frame bridge is explained by the sketches on pp. 134–5. In the figure on p. 134 the distribution of bending moment produced by uniform loading on a simple beam is shown. If two cantilevers are added the addition of loads at the end (ignoring the self-weight of the cantilevers) will produce the distribution of bending moment shown. The addition of the bending moments for the uniform loading and the point loading will result in the effect shown (on p. 134). When a portal frame is used, the load is produced by horizontal reactions at the foundations. The foundations must therefore be designed to resist horizontal forces as well as the vertical forces. The ratio of the beam length to the column height will generally be governed by shipping requirements. However, the distribution of moment can be altered by varying the size of the beam and columns. If the column is slender and the beam deep, the moment distribution will approach that of a simple beam. If the beam is slender compared with the column the moment distribution will approach that of a beam held rigidly at both ends. The same applies if the column member is replaced by compression and tension members (strut and tie).

Next, before we consider the application of prestressing to portal frame bridges, it is necessary to take a brief look at its development

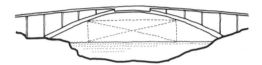

The construction of an arch bridge can involve considerable wastage of material as navigational requirements necessitate shipping lanes of constant height.

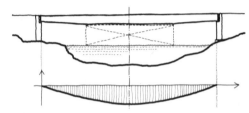

If a large opening is spanned by a simple beam its depth may be excessive.

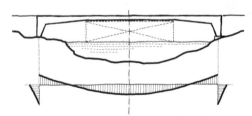

By enclosing the opening with a rectangular (portal) frame a shallower beam depth can be achieved as the bending moments are reduced.

When the Esbly Bridge was completed the technique of prestressing concrete was virtually unknown in many countries. A concrete bridge of these proportions could not have been built before the development of prestressing techniques.

The distribution of bending moment with uniform loading on a simple beam.

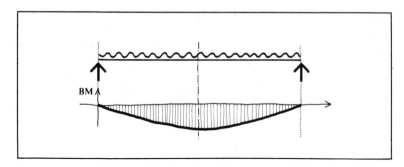

The addition of cantilevers with loads at each end will reduce the moment at the centre of the beam.

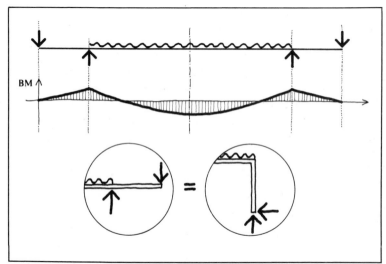

in general and the work of Eugène Freyssinet in particular. Prestressing aims to induce a state of stress in a structure so that in all loading conditions the stress at any section is within the limits of the material used. This technique is particularly suited to concrete, which is weak in tension. A concrete section can be subjected to an initial compressive stress that will eliminate tensile stresses caused by the loading. Such a situation is illustrated on p. 136, top. The simplest way of achieving this is to subject the concrete to an axial compressive force. But it is more logical to apply the force as shown on p. 136, centre. This will produce the maximum compressive force at the zone of maximum tension—that is, in the bottom fibres of the beam. The eccentricity of the prestress force can be varied along the length of the beam so that the tensile stress is exactly balanced at each section. This means that the line of action of the prestress force will be parabolic. Various methods of achieving the necessary prestress are shown on p. 138. Prestressing also means that the member can be made up of a number of segments or voussoirs, which are later stressed together. Thus the prestress force is equivalent to the thrust that holds the voussoirs together in a masonry arch.

Prestressing was applied to many early steel trusses and a theoretical solution to the problem was developed by Alberto Castigliano (see p. 173). Here truss members were extended by means of a turnbuckle, so that stresses were introduced before the load was applied. These stresses were of course opposite to those produced by the applied load.

Although concrete had been in use for many structures from the latter part of the 19th century, the majority of engineers had no real understanding of the nature of the material. Notable exceptions were Robert Maillart (see p. 125) and Freyssinet. Eugène Freyssinet was born in 1879 of peasant stock. He considered that a firm foundation to his technical education was provided by the self-reliance inherited from his ancestors—the ability to arrive at the most practical solutions to the everyday problems of peasant life. On leaving a Paris council school, Freyssinet was offered a place at the Ecole des Ponts et Chaussées. Even so, Freyssinet did not acquire the approach of the typical Polytechnician—that is, complete faith in the power of mathematics. He has said: 'For me only two sources of information count: direct perception of the facts and intuition, in which I see the expression and the beauty of mathematics. . . . Intuition must, of course, be governed by experience but whenever it is in contradiction with the results given by calculation, I renew my calculation and in the end, my collaborators always affirm that it is the calculation which is at fault. . . . Nor do I contest the utility of mathematics in our profession—I have had to resort to them on occasions. . . .'

Freyssinet was disturbed by the illogical behaviour of reinforced concrete, in which the steel and concrete are subjected to tensile strains so that cracking occurs. In 1904 he thought up an idea for stresses and strains. To quote again from his theories: 'In inventing this word [prestress], I also defined its sense: the forces of prestress are the forces created by the designer to provoke the stresses before, or at the same time as, those resulting from the loads and which, combining with these, give, at every point, resultant stresses which

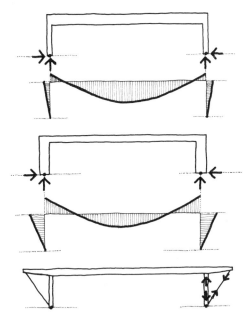

The distribution of moment can be varied by altering the proportions of the members. A deep beam and slender column will produce little reduction in the moment at the centre of the beam (top). If the beam is slender compared with the column, the moment at the centre of the beam will be considerably reduced (centre). The column may be replaced by a compression and tension member (bottom).

A concrete beam subjected to a uniform compression which eliminates tension due to bending.

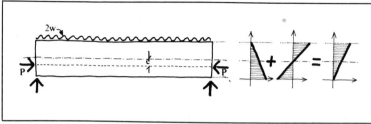

prestress + loading stress = total stress

The compressive force is applied eccentrically to produce maximum compression at the bottom of the beam.

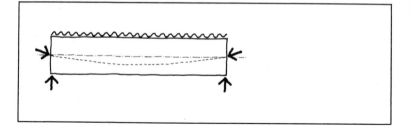

The eccentricity of the prestress force can be varied along the length of the beam.

are limited to values that can be sustained indefinitely by the materials without the latter suffering any change.'

It was Freyssinet's knowledge of the deformation of concrete under load that enabled him to build long-span concrete bridges between 1920 and 1930. An understanding of the phenomena of shrinkage and creep in concrete and the resulting deformations was essential for the successful construction of arches and prestressed concrete members. All early attempts at prestressing failed because the effects of creep and shrinkage were not allowed for. Assume that a steel bar is initially tensioned up to a stress of 20,000 pounds per square inch and anchored at each end of the unit, thus producing a stress of 1000 pounds per square inch in the concrete. This stress will shorten the member by an amount x. If the prestress is applied a few days after the concrete has been cast there will be a further reduction in the length of the member as the concrete gradually

Annet Bridge, one of five similar bridges constructed by Freyssinet.

drys out—that is, as the concrete shrinks. Let this value be y. Further, the prestress force acting over a long period of time tends to squeeze water out of the pores of the material with an additional gradual reduction in the length of the member. Let this value be z. After a long period of time equilibrium is reached after which there is no further reduction in length and the total reduction is xyz. A knowledge of the properties of the concrete and steel enables these values to be expressed numerically with the result that the final prestress force is zero. It should be noted that the loss in prestress is gradual since creep and shrinkage are time-dependent phenomena.

In the 1920s Freyssinet conducted some experiments with concrete prestressed with piano wire. This wire could be tensioned up to a very high stress so that the resulting loss would only be a small percentage of the initial stress. This produces the same percentage of stress reduction in the concrete. From this it is obvious that it

137

Pretensioning: the wires are tensioned between anchorages, the concrete is poured and allowed to mature, and then the wire anchorages are released.

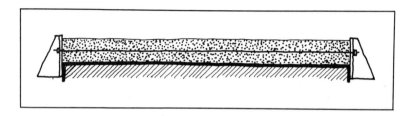

Post-tensioning: the concrete member is cast with a duct running through it, cables are threaded through the duct and the concrete stressed by jacking against the ends of the beam.

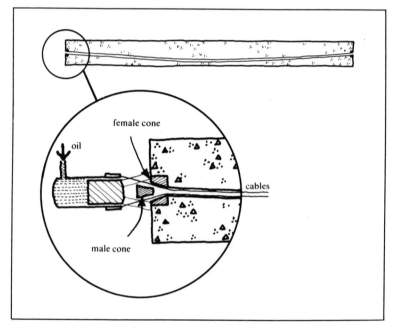

is necessary to use high-tensile steel in prestressed concrete, and that this loss factor has to be taken into account when the member is designed if tension is to be avoided.

The use of high-tensile wire for prestressed work means that it is possible to create the required prestress force with a relatively small area of steel. This was one of the reasons for the rapid development of prestressed concrete after the Second World War when steel was in short supply. In France in particular the pioneer work of Freyssinet led to the large-scale use of prestressed concrete for bridge structures in the immediate post-war period. Especially notable are the portal-frame bridges constructed by Freyssinet over the River Marne at Luzancy, Esbly, Annet, Tribardou, Changis and Ussey.

The Luzancy Bridge, with a span of 180 feet, was started in 1941 and completed after the war in 1945–6. The remaining five, all with a span of 240 feet, were also identical in form and constructed between 1948 and 1950. The application of prestressing to portal frames poses further problems. When a prestress force is applied to a simply supported beam, it is assumed there will be no restraint to the longitudinal compression of the beam caused by the prestress, since one end will be on a sliding or roller bearing. But if a prestress force is applied to a portal frame, some restraint will be imposed on the longitudinal compression as a result of the frame legs being fixed in position. This produces a horizontal reaction at the column feet and introduces bending moment that will induce tension in

The slender Esbly Bridge undergoing test loading.

139

Fabrication of the I-section voussoirs.

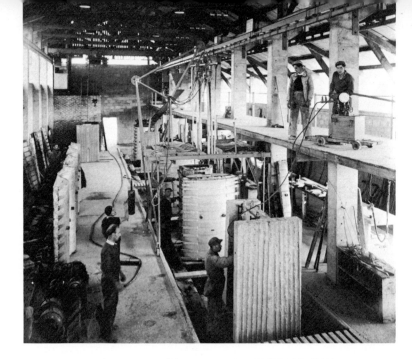

The I-sections were stressed together in groups, and two sets of voussoirs were placed side-by-side.

There is no restraint to compressive shortening as one end of the beam is on a roller.

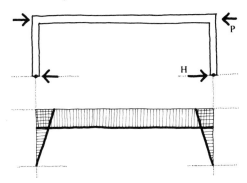

A portal frame provides restraint to shortening of the beam, which induces additional moments in the structure.

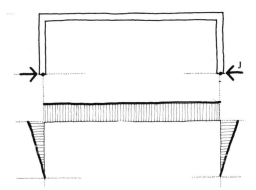

The frame can be jacked (J) to produce a moment distribution to compensate for that shown above.

Vertical prestressing (right) was used to prevent diagonal cracking in the thin webs.

the bottom of the beam at midspan. Further shortening of the beam brought about by creep and shrinkage will increase this tension. To compensate for creep and shrinkage, Freyssinet introduced jacks in the Esbly Bridge.

The profile of the bridge was influenced by navigational head-room requirements and the deck construction was made as light as possible to reduce the thrust on the abutments. The thrust was taken up by horizontal concrete frames placed in the ground behind the abutment walls. Since five identical bridges were being constructed, a great deal of mechanisation was possible. The main beams were constructed from prefabricated voussoirs of I-section; they were cast in two stages. First the top and bottom flanges and gussets were cast and steam-hardened round loops of high-tensile steel wire anchored in both flanges. This steel was then tensioned by jacking the flanges apart, and the web concrete was cast round the wire loops. When the web concrete had been steam-cured the jack force was removed, thus prestressing the webs. This was necessary since the webs were only four inches thick and would be subjected to high shear stresses. Shear induces diagonal cracks in the webs. This is prevented if the webs are prestressed vertically.

The prefabrication of the voussoirs also meant that much of the creep and shrinkage had already taken place when the units were prestressed together, thus minimising the shortening effect described earlier. The voussoirs were prestressed together in groups of six, after they had been connected by *in situ* concrete joints. Since it is almost impossible to match faces in concrete work, precast units generally have to be connected by *in situ* joints to avoid local concentrations of stress from the uneven contact surfaces. Two sets of voussoirs were placed side by side, the joints made as before and the lengths stressed together. This improved stability during handling.

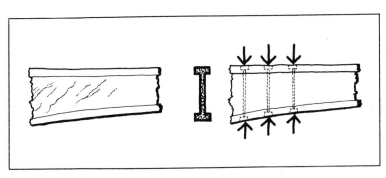

The units were hoisted into position in three sections, using a guyed mast and block and tackle. The portal legs were placed first, then the intermediate and central sections. Each section was connected by an *in sutu* concrete joint and then tied in by prestressing cables. The majority of the cables ran along the top of the beam units and were then covered in concrete. A smaller number of cables were passed through ducts prepared during prefabrication and then tensioned. These cables passed along the bottom flange over the central portion of the span. This rather neat cable arrangement was made possible by the increase in corner moment produced by the jacking. A very shallow midspan section could thus be adopted.

The design and construction of this incredibly slender bridge— the depth of construction at midspan being just over three feet —demanded a profound knowledge of the nature of concrete as a structural material. It would not have been possible to build such a bridge in concrete before the development of prestressing techniques. When the Marne bridges were constructed, prestressed concrete was virtually unknown in many countries and many engineers considered that it had no future. In his designs for the Marne bridges, Freyssinet, the father of prestressing, demonstrated the potential of prestressed concrete for bridge building and his ability to apply an abstract idea to a particular branch of structural engineering. Within 15 years of the completion of the Marne bridges, prestressed concrete had become the most widely used material for bridge construction in several countries.

Construction in progress. The units were hoisted into position in three sections.

The three sections of the frame completed.

Maracaibo Lake Bridge

Industrial growth and the rapid increase in car and truck ownership in Europe and America has led to the development of integrated highway systems. Such systems involve the construction of bridges notable (among other things) for their exceptional length. Typical examples are the Oosterschelde Bridge in Holland, the Chesapeake Bay bridge-tunnel in the U.S.A. and the bridge spanning the Maracaibo Lake in Venezuela. The Oosterschelde Bridge, over three miles long, joins two islands in the Eastern Schelde to form a road link connecting a peninsula and the islands to the mainland, thus reducing the journey from Rotterdam to the extreme south-west of Holland by about 30 miles. This bridge consists of 48 spans of 312 feet in prestressed concrete.

The Chesapeake Bridge stretches $17\frac{1}{2}$ miles across the lower Chesapeake Bay, connecting the lower tip of Virginia's eastern shore with the southern Virginia mainland. The bridge-tunnel has four basic components: 12 miles of low-level concrete trestle, two 1-mile tunnels, two steel bridges and four man-made islands. A motorist using this bridge saves $1\frac{1}{2}$ hours of driving time. The Maracaibo Bridge, over 5 miles in length, links the oil country on the eastern shore of the lake with Maracaibo, the capital of the Venezuelan state of Zulia. The five 780-foot spans of the Maracaibo Bridge, constructed in prestressed concrete, dwarf those of the Oosterschelde and Chesapeake bridges. A common feature of all these bridges is the use of reinforced and prestressed concrete and extensive prefabrication of the structural elements in order to cut down construction time.

Economic development on Venezuela made it necessary to improve existing communications, and the ferries plying between Maracaibo and the eastern shore of the lake had become incapable of handling the rapidly increasing volume of traffic. The Venezuelan government therefore decided to investigate the possibility of crossing the lake by a bridge or a tunnel. Five alternative routes were proposed and examined; the route finally chosen was picked because it dovetailed most satisfactorily into Maracaibo's traffic system. In 1956 ten firms were invited to tender for the construction of a bridge or tunnel or a combination of both. The specification stipulated four highway lanes, a single track railway and openings to provide

The Oosterschelde Bridge, completed in 1966, has 48 spans of 312 feet.

The structure of the Maracaibo Bridge incorporated vertical V and H trestle piers. For the main spans, tied cantilever beams were adopted.

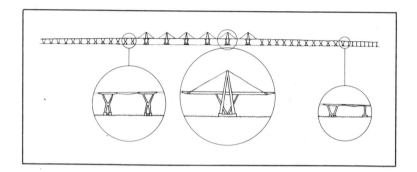

adequate clearance for shipping. However, because of later stipulations by the shipping and highway authorities, these first tenders were abandoned and new tenders were invited in 1957.

The second specification asked for a central clear span of 1300 feet flanked on each side by five clear spans of 490 feet. The headroom required for these spans was about 150 feet. Twelve tenders in all were submitted, eleven with a steelwork superstructure and one of reinforced and prestressed concrete. The Venezuelan government commission recommended the acceptance of the concrete structure, which was based on a design by Professor Riccardo Morandi. This tender was submitted by Messrs Precomprimido, C.A. & Julius Berger, A.G. and four reasons were given for its acceptance. First, the use of concrete structure greatly reduces maintenance costs. (A steel structure of this length needs continuous painting.) The second reason was aesthetic—the visual appeal of the design—and the third was economic because less foreign exchange would be spent on imported materials. Lastly, it was thought that Venezuelan engineers would gain experience in the use of prestressed concrete. In 1958 the design was amended by replacing the 1300-foot span originally planned with five main spans of 780 feet. Work on the bridge began in April 1959.

The structure is laid out in the following way. Starting from the eastern shore of the lake there is a 1300-foot embankment, followed by 20 short spans of about 120 feet. This length of span had been fixed for the road-and-railway bridge and had to be kept, since the foundations were already complete when it was decided the bridge should be for road traffic only. The next section comprised 77 spans of about 150 feet with a slightly increased gradient. The length of the following spans was increased to about 275 feet, and the vertical piers replaced by V and H trestle piers. For the five main spans (those of 780 feet) tied cantilever beams were adopted.

Maracaibo is often affected by shock waves from earthquakes in neighbouring areas, and the specification stipulated that should one span become damaged in this way the adjacent spans would remain intact. This necessitated the adoption of a simple beam system in which each span is independent of the next. The structural principle involved is illustrated at right. The 120-foot and 150-foot spans are of similar form, consisting of prestressed-concrete fish-belly beams supported on vertical piers. The beams have a fixed bearing at one end and one supported on steel-sheathed concrete rollers at the other, thus allowing for movement caused by thermal expansion or contraction and creep and shrinkage of the concrete. The deck consists of four longitudinal beams connected by a top slab and transverse members (diaphragms at quarter-span points). This ensures that a heavy wheel load moving along the line of one beam is distributed on to the adjacent beams; such a load distribution allows the structural depth to be minimised. The 275-foot spans consist of double cantilever beams above the trestle piers with simply supported 150-foot span fish-belly beams placed between. These form the approach spans for the five main spans of 780 feet over the navigational channel. The 780-foot spans consist of a series of double cantilever beams, each pair of cantilevers of an overall length of 616 feet connected by a simply supported 150-foot span fish-belly beam. To reduce the structural depth of the box-section cantilever beams to just over 16 feet they were supported by inclined ropes connected about 47 feet from each end. The inclined ropes are suspended from the top of 300-foot-high towers. The cantilever arms are thus propped by the vertical component of the rope tension while the horizontal component helps to prestress the deck. To ensure that the horizontal component of the rope tension was distributed evenly over the entire width of the deck the connection was made by prestressed transverse beam $19\frac{1}{2}$ feet by $6\frac{1}{2}$ feet. Because of the relatively flat slope of the ropes, the axial force induced in the deck eliminated the need for any but nominal reinforcement in most parts of the deck. The use of the structure itself to provide the prestress is typical of Morandi's quest for coherence in his work. The beauty of this bridge derives from the adaptation of a well-established structural principle, the propped cantilever, to a large structure with the greatest economy and simplicity. It is interesting to compare this structure with the strikingly similar Forth Bridge (1890), the influence of which Professor Morandi acknowledged.

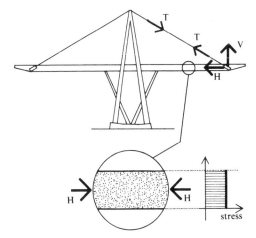

The structural principle of the main spans: the cantilever is propped by the vertical component of the tie force and prestressed by the horizontal component.

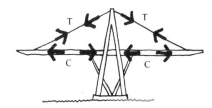

A comparison of the Forth and Maracaibo bridges in terms of structural concept.

The Chesapeake Bay Bridge has an overall length of 17½ miles including 12 miles of low-level concrete trestle construction and two mile-long tunnels.

Construction of the five-mile Maracaibo Lake Bridge was completed in 40 months.

The 275-foot spans consist of double cantilever beams above the trestle piers with 150-foot-span fish-belly beams placed between.

The Maracaibo Bridge is a vast technical and constructional achievement and could not have been accomplished by one man. From the planning stage to construction a large number of Venezuelan and European engineers worked jointly in Wiesbaden, Caracas, Rome, Maracaibo, Zürich, Paris and Lisbon. During construction 2630 men were employed on the project, which was completed within the contract time of 40 months. Professor Morandi's design served as a basis for the structural analysis and detail plans, which were prepared jointly by the contracting consortium and Professor Morandi. The complex aspects of the geological investigations were handled by Professor Kérisel, Paris. The National Laboratory of Portugal, Lisbon, was commissioned by the Venezuelan government to carry out model tests of the structure and the examination of the structural analysis and working drawings, and the control of the design was entrusted by the Venezuelan government to Professor P. Lardy, G. Schnitter and Dr F. Stüssi, of the Swiss Federal Polytechnic, Zürich. Firms associated with the contracting consortium included Messrs Grün & Bilfinger A.G., Mannheim, Philipp Holzmann A.G., and Wayss & Freitag K.G., both of Frankfurt. The bridge was truly an international venture.

To keep within the stipulated contract time, large-scale site organisation and prefabrication of the structural elements, formwork for the concrete and reinforcement were necessary. The site plant was erected on the west side of the lake near Punta Piedras, a few miles from the centre of Maracaibo. An area of about 300,000 square yards was prepared, together with miles of road to give access to the plant in the rainy season. Maracaibo is situated in a tropical climatic belt and the annual rainfall of 22 inches is produced by twenty to thirty violent cloudbursts. In the winter there are long windy periods, which make construction work difficult. Most of the site was taken up by the plant required for prefabricating the concrete units, which involved handling loads of up to 200 tons. Construction work was carried out day and night; accommodation was therefore provided for the construction workers and their families at a camp near the plant site. Work began with the erection of drilling rigs on platforms in the lake. These were put up to determine the nature of the soil below the lake bottom and hence the type of foundation to be adopted. Tests carried out by the soils consultant showed that the lake bottom consisted of a layer of soft mud, below which was a firm stratum of slightly clay-like and silt-like sand. Conventional driven or cast-in-the-ground piles were adopted for

Construction of the piers in progress.

the shorter spans, but for the heavier loads a new type of pile was developed. These piles were subjected to extensive test loading without failure with loads of more than 2000 tons. For the actual structure design, pile loads of 500 to 1000 tons were adopted. Once the pile groups had been driven, it was necessary to construct a cap over the top so that construction of the bridge piers could be begun. Wave action made *in situ* erection of the shuttering for the concrete pile impossible. The shuttering units were therefore erected on the shore, placed on barges and towed out to the pile groups. The shuttering sets for the piers together with the reinforcing cages were erected on the shore and then placed in position by floating cranes. The floating cranes also transferred the concrete from floating mixing plants to the steel shuttering.

Over 70 per cent of the length of the bridge consists of prefabricated girders and wide use was made of standardisation in order to cut down construction time. The 120-foot spans were bridged directly with 80 girders and the 150-foot spans with 308 girders. For longer spans, the 140-foot girders were placed between the cantilever sections. These girders were cast on the shore and floated out by barges. The 120-foot-span girders were raised to the necessary level by a hydraulic mechanism incorporated in the barges. The 150-foot-span girders were raised by a 250-ton Ajax floating crane, since they were to be seated on higher supports.

The pile caps for the 275-foot spans were constructed in the same way as they were for the 150-foot spans—that is, the shuttering was fabricated on the shore and placed by floating crane on the pile heads. The V and H piers were connected by using climbing shuttering; every possible use was made of standard components. The double cantilever beams at the top of these piers were concreted *in situ*, making use of steel service girders. Much of the preliminary work was carried out on the shore, including prefabrication of the shuttering and the reinforcing cages. The steel service girders were positioned by floating cranes and then removed when the concreting had been completed.

The pile caps for the 780-foot spans rested on 62 piles. The pile caps had a plan area of about 127 feet by 112 feet and were up to 19 feet in depth. Two tower cranes were mounted on each pile cap to ease the erection of the X frames supporting the deck and the towers. Materials for the towers had to be lifted up to 300 feet and

Placing the girders for the shorter spans: the 250-ton Ajax floating crane is carrying a 150-foot-span girder.

The steel service girder carrying the prefabricated shuttering, opposite, being raised into position.

152

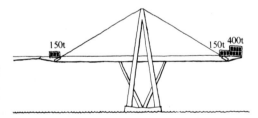

150t   150t 400t

*As the suspended beams were placed in position
the concrete blocks were gradually removed
until 150 tons of blocks were left. This load was
gradually reduced as surfacing progressed.*

for the cantilever moved up to 300 feet horizontally.

To cast the concrete for the 236-foot cantilever arms, steel service girders were used. These girders were supported at one end from the completed pier cap and from the other by temporary steel piers supported on an auxiliary foundation. The cantilever arms of the triple box section, 16 feet deep with 10-inch-thick walls, were concreted in six sections. The concrete was produced in a floating mixing plant, which was anchored alongside the pier. It was then carried to the tower crane for pouring by buckets. At this stage the cantilever arm was supported by the steel service girder. By tensioning the ropes, the load was gradually removed from the temporary piers. When the final rope was tensioned the piers were completely relieved of load. The horizontal component of the inclined rope force then acted as a horizontal thrust on the cantilever arm. Later, the service girders were lowered, thus breaking the bond between the concrete and the girder. The temporary piers were then lifted and secured to the service girders and the service girder was placed on a twin by the Ajax floating crane at one end and winches installed on the pier cap at the other. The removal of the service girders reduced the rope forces by 1000 tons. To avoid redistributing the forces in the partially unloaded system, concrete blocks weighing 550 tons were placed on each end of the cantilever arms. Before placing the suspension girders an appropriate number of blocks were removed. To allow for the weight of the surfacing and so on, 150 tons was left. This load was removed as the surfacing progressed.

Each cantilever arm is supported by two sets of ropes forming a 32-rope system. Each system carries a permanent load of 5425 tons or 170 tons per rope. The traffic loading produces a further load of 14 tons per rope. It can be seen from these figures that the self-weight of the structure dominates the live load, which produces little more than eight per cent increase in rope load.

This bridge is an outstanding example of large-scale construction in prestressed concrete; it has been followed by two bridges, both designed by Professor Morandi, using the same structural principle. The first is the Polcevera Creek Bridge, which forms part of the Genoa–Savona Autostrada, and the second a recent project for a bridge over Göteborg Harbour in Sweden. This bridge has a proposed central span of 1350 feet. In these two schemes the structural system has been somewhat simplified and the inclined ropes are encased in prestressed concrete shells for permanent protection.

Opposite, lowering the service girder for the cantilever arms.

The service girders for the main spans have been removed and the fish-belly beams are being placed between the ends of the cantilever arms.

154

Norderelbe Bridge

The Norderelbe Bridge which carries the southern bypass motorway in Hamburg, is typical of many cable-stayed bridges constructed, mainly in Germany, in the 1950s and 1960s. Possibly the cable-stayed bridge will soon become established as a classical bridge type in the same way as the arch or suspension bridge. It forms a useful transition between the continuous girder and the suspension cable; it is therefore particularly suited to spans in the range of 500 to 1500 feet. Like the Norderelbe Bridge, most cable-stayed bridges have been constructed with a steel deck but there are some examples in concrete, such as the famous Maracaibo Lake Bridge.

With spans of more than 500 feet, continuous girder bridges, constructed either in steel or concrete, need a large structural depth often of over 30 feet. The world's longest steel-girder bridge is the highway bridge over the River Sava at Belgrade. This is a three-span continuous girder structure with a main span of 856 feet and two side spans of 246 feet. The structural depth at the intermediate piers is about 32 feet. The use of cables to support the deck at these intermediate positions would mean that the structural depth could be reduced to less than half this value. Suspension bridges have been used for spans less than 1500 feet in the post-war period, but with the increase in heavy vehicular loading, the ratio of permanent load to vehicular load is low and poses a difficult problem of stability. The cable-stayed girder is therefore a satisfactory compromise between the continuous girder and the suspension cable.

In such a bridge, the intermediate piers found in a continuous girder bridge are replaced by cables above the deck. This form of construction results in little or no interference to shipping (if any) and also reduces to a minimum the amount of construction work required in the water itself. The deck can be supported by a number of cables, which means that the distance between points of support for the deck girders will be small, resulting in reduced structural depth.

In general the deflection of the intermediate piers supporting a continuous girder is small; therefore the girder will not deflect at the points of support. If the piers are replaced by inclined cables there may be a considerable extension of the cables so that the girder does deflect at these points. If the extension of the inclined cables is substantial, the girder will tend to act as a simple beam between the

An aerial view of the Norderelbe Bridge, showing the principal elements – towers, cables, and deck.

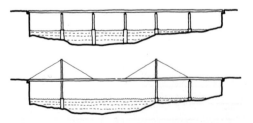

The intermediate piers supporting a continuous girder can be replaced by cables above the deck, resulting in a much larger clear navigational width.

The world's longest steel girder bridge, which spans the River Sava at Belgrade. The main span is 856 feet and the structural depth at the intermediate piers is about 32 feet.

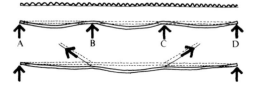

This sketch shows how a highly extensible cable may result in the girder acting as a beam between simple end supports due to the deflection at points B and C.

end supports. This will produce much greater stresses and so an increased structural depth. However, it is possible to minimise this effect by pre-stretching the cables, so that the structure behaves more like a continuous beam on rigid supports. (The design of the cables, a vital structural element, will be discussed more fully in the latter part of this chapter.)

Although the cable-stayed bridge is still comparatively rare, the idea of using cables or stays to support a girder or mast is not new. One of the first examples was that proposed by C. J. Loscher in 1784. Other early examples were the designs of Poyet (1821) in France and Hatley, an Englishman, who designed a harp-shaped suspension system (1840). The most notable instance of the use of stays to support a bridge deck in the 19th century is the Albert Bridge over the River Thames. This bridge was completed in 1873 and is really a suspension bridge stay reinforced with wrought-iron flats.

Early attempts to use a girder stayed by tension members often suffered from the disability mentioned earlier—that is, the extension of the stays. The tension members were made of timber, round bars or chains and were often connected when slack. Consequently they provided little support to the main girders and overstress resulted. Recent advances in analytical techniques and the use of more sophisticated cable systems have enabled the problems met in these

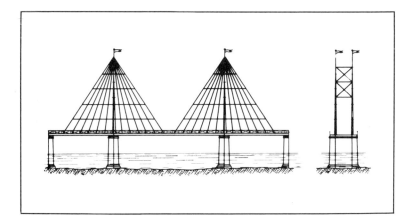

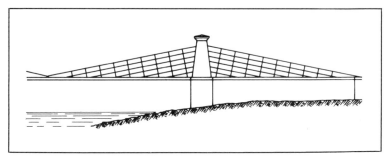

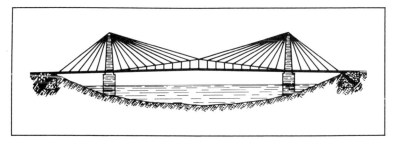

Early examples of designs for cable-stayed bridges.

earlier structures to be overcome. This was largely the work of German engineers and in 1956 the first large cable-stayed bridge was designed and built across the Strömsund in Sweden by a German firm. It has a main span of just under 600 feet and towers built as trapezoidal frames, the bridge having a two-plane cable system. The two-plane system is a common feature of the earlier cable-stayed bridges, the cables appearing to intersect if the bridge is viewed at an angle. Aesthetically this is not very attractive, and in later

159

bridges such as the Norderelbe a single-plane system is used—in other words, all the cables are in the same plane. The cable systems and towers in the Strömsund and Norderelbe bridges are illustrated at right, top. As can be seen on p. 163, if the Norderelbe Bridge is viewed from one side, there is no intersection of the line of cables and a motorist using the bridge has an uninterrupted view on one side. This was the first bridge to be built with single-plane cables and was completed in 1962.

In a two-plane system, the cables can be anchored in the deck either inside or outside the handrailing. If they are anchored outside, the loads have to be transmitted from some distance outside into the main deck structure, as in the Maracaibo Lake Bridge. But alternatively, if the cables are anchored to the deck within the handrailing, two additional widths of bridge deck are required for the towers, and the cables, and the width of the deck has to be increased. In a single-plane system, the anchorage point for the cables can be in the central reservation dividing the carriageways, with the result that no extra deck width is necessary. At first sight a bridge with a single-plane cable system may look rather unstable. If a line of heavy traffic moves over the bridge close to the outer edge of a carriageway there will be considerable twisting. To avoid excessive distortion of the deck structure, its form must be a kind resistant to high twisting moments. The Norderelbe Bridge has a central box girder, which provides the resistance to the twisting induced by vehicles crossing the bridge. Say, for example, that the centre of gravity of a car is about 30 feet from the centre line of the deck. The twisting effect induced will be related to the vehicle load and the distance of its centre of gravity from the bridge centre line. The central box girder must be designed to resist the twisting action induced by the vehicles and will need two point supports at one end.

A most important structural element is of course the cable itself. The cables used for the Norderelbe Bridge are of the locked-coil type. This type of cable has several advantages over the conventional spiral rope manufactured from round wires, which is used for general engineering machinery such as cranes and hoists. Since the cable is a tension member, its diameter can be relatively small compared with a compression member, which must be proportioned to avoid buckling failure. Further, the diameter of a locked-coil cable can be less than that of a spiral rope, because the density of material in an area of given diameter is about 30 per cent more for the locked-coil type. This greater density is produced by the larger

The Albert Bridge over the Thames, completed in 1873, is a suspension bridge stay-assisted with wrought-iron flats.

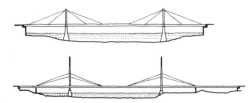

An illustration of the cable systems and towers used for the Strömsund and Norderelbe bridges. The latter uses a single-plane cable system.

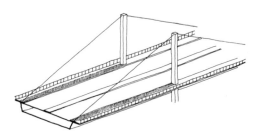

Cables anchored within the handrailing, making two additional widths of bridge deck necessary.

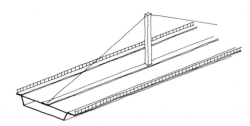

In a single-plane system, the anchorage point for the cables can be placed within the central reservation, and no additional deck width is necessary.

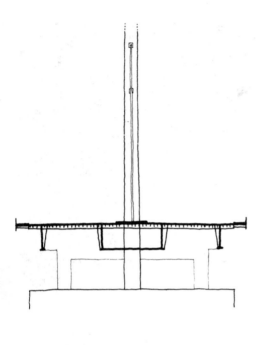

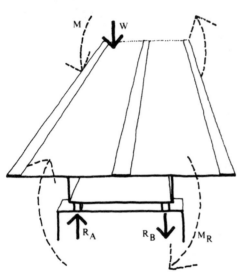

The Norderelbe Bridge has a central box girder which provides resistance to the twisting induced by eccentric vehicle loads.

contact area between the adjacent wires. Further advantages of the locked-coil cable are greater protection against corrosion and increased resistance to extension. This second quality is important, because, as mentioned earlier, large extensions of the cables under load will increase the deck-girder moments and require a greater depth to the bridge deck. It has been shown that the resistance to extension of a locked-coil cable is related to its length and the stress. For the Norderelbe Bridge the cable length is such that its resistance to extension is not much less than that of solid steel.

The constant movement of vehicles over the bridge deck will cause variations in the cable loads, which may lead to failure because of fatigue. Repetitions of stress, especially when combined with local defects in a metal and sudden loading, can lead to a progressive deterioration. In such circumstances, failure can occur when stress is numerically less than that produced by single load.

In addition the cables are of course subjected to a permanent stress produced by the self-weight of the deck construction. Indeed in a large-span bridge the vehicular loading may be a relatively small proportion of the total load. The stress variations may therefore not be very severe. To ensure an adequate margin of safety the allowable loads on cables are restricted to allow for loss in strength from fatigue.

Next we must consider how the loads in the cables are transmitted to the deck and the towers. The Norderelbe Bridge has single box-section towers 172 feet high above the deck. The cables are attached to the towers at heights of about 57 feet and 75 feet above the deck. The part of the tower above the top cable has no structural significance but is intended as a tribute to Hamburg's city fathers. Each cable consists of ten ropes, which are assembled in two layers of five, clamped together and formed into a rectangle. The top cables are permanently fixed to the towers whereas the bottom cables are supported on movable bearings.

At the deck level, the ropes are held in a clamp, from which they fan out to their anchorage points in the central girder. The transmission of the large cable force into the deck structure is one of the major design problems in cable-stayed structures. It is far from easy to analyse the local effects of a highly concentrated force and the reason for fanning out the ropes is to increase the area over which the load is distributed and so reduce the stress concentration. It is also necessary to ensure that the cable force is transmitted over the full width of the deck, which acts as a large beam. The means of

anchoring the ropes in the deck structure is shown above. The deck structure for the Norderelbe Bridge consists of the central box-section girder (described earlier), in which the cables are anchored, and two outer girders of L-section. The three main girders are connected by means of cross-girders and the steel deck. The cross-girders are located at about 70-foot intervals and the steel deck acts as a kind of a stiffened plate. The stiffened plate is now a common feature in bridge structures, though at one time it was more commonly associated with aircraft and ship structures. There is now a striking similarity between the fuselage structure of an aircraft and the box-deck structure of a steel bridge. Steel is a comparatively heavy and expensive material, and as the need for long-span structures increases, weight is becoming a much more important factor in the efficient design of bridges such as the Norderelbe. As a load-carrying member, a thin steel sheet would undergo considerable deformation even if supported at closely spaced intervals. By adding stiffeners the resistance to deformation can be increased considerably with very little added weight. This principle was used in the design of the steel deck for the Norderelbe Bridge. The overall depth of the bridge deck is about 10 feet and the weight of the bridge is about 66 pounds per square foot of deck area. This is equal to the weight per

At deck level, the cables are held in a clamp from which they fan out to their anchorage points in the central girder.

The cables are attached to the towers at heights of about 57 and 75 feet above the deck. The part of the tower above the top cable has no structural significance.

163

square foot of a solid slab of concrete about 5½ inches deep. Obviously then a steel deck structure of the form used in the Norderelbe Bridge would be much lighter than a concrete deck structure designed to span the same distance and carry the same vehicular loading. There are of course other factors besides weight involved in considering the merits of a particular scheme such as availability and cost of basic materials, method of construction and maintenance. However, as understanding of the properties of materials increases and electronic computing techniques and construction methods develop, minimum-weight design will gain in significance, with the result that there will be a much closer relationship between aircraft and bridge structures.

The general arrangement of the structural elements of the bridge can be seen in the picture on p. 157. The construction method adopted was the cantilever principle, the same procedure used by Baker and Fowler on the Forth Bridge. The first stage was to erect the side spans and the towers. The main span was then cantilevered out to a distance of about 185 feet from each tower. At this stage the deck structure acted as a free cantilever, which is the condition in which maximum bending moment and deflection occur. (In the design of any structure care must be taken to check the stresses arising during construction as these could easily be much greater than those in the completed structure. It may be possible, however, to allow a temporary increase in stress during erection since the erection loads will not be permanent.) With a cantilever of 185 feet it was then necessary to support the deck with temporary guy ropes to limit the end deflection as more bays were added. The anchorage points for the cables are at a distance of about 208 feet from the towers and when the anchorages had been built into the central girder, the cables were connected and tensioned. Thus the structure was completed without making use of a temporary support in the river. It is interesting to compare the erection procedure for the Norderelbe Bridge with that for the Gladesville arch bridge in Australia.

The construction of the arch required the use of elaborate staging to support the voussoirs during erection. The Norderelbe Bridge is an excellent example of modern steel bridge design developed from ideas conceived and practised by 19th-century engineers. Three important features of this structure—the use of a box girder, the cable stays and the cantilever method of construction—had been used in the 19th century, but not all three on one construction. The

A view of the partially completed deck structure showing the central box section, the two outer girders, the cross girders, and the stiffened steel-plate deck.

The Norderelbe Bridge is typical of many cable-stayed bridges constructed, mainly in Germany, during the 1950s and '60s. It may soon become established as a classic bridge type.

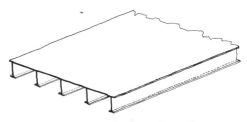

A thin steel sheet can be strengthened by means of stiffeners.   165

The first stage in construction was to erect the side spans and the towers.

The main span was cantilevered out to a distance of about 185 feet from each tower.

The centre bay of the bridge, completing the deck structure, was erected without making use of temporary supports in the river.

166

pioneer work in the use of box girders was of course the Britannia Tubular Bridge designed by Robert Stephenson and completed in 1850. But though developed from the Britannia Bridge, the sophisticated form of the Norderelbe Bridge deck structure is a product of 20th-century technology. As mentioned earlier, the development of the cable-stayed bridge since 1950 is attributable largely to the enterprise of German engineers, but examples are now appearing in other countries. The first major bridge of this type to be built in Great Britain was the Usk Bridge at Newport, Monmouthshire. This structure has a main span of 500 feet, its box-section deck structure supported by 12 locked-coil cables placed outside the main carriageway. The footways cantilever beyond the cable line. The four towers supporting the cables are of reinforced-concrete construction and the bridge was completed in 1962. In 1966 a bridge over the River Wye, close to the Severn Suspension Bridge, was completed with a main span of 835 feet. This structure has a similar form to the Norderelbe Bridge, with single support towers and cables running in line with the central reservation. It differs from the German bridge in two ways: the deck is a multi-cell box structure and there is only one set of cables running from each tower. On a smaller scale, the cable-stayed girder principle has been applied to the construction of footbridges. A typical example of a footbridge constructed in Germany is shown here; it is of very distinctive and pleasing form.

As these examples show, the cable-stayed bridge is playing an important part in the development of the art of bridge building, although its potential has yet to be fully exploited for both steel and concrete structures.

An example of the use of the cable-stayed principle: a footbridge crossing a motorway in Germany.

Gladesville Bridge

The erection of the Gladesville Bridge in Sydney, Australia, illustrates how a method of construction widely used during the Roman period can still be economical in the latter half of the 20th century. The bridge is a voussoir arch of hollow concrete elements, built in the same manner as early stone bridges. As we have seen, the principle of the arch was known well before Roman times. Neolithic man discovered that an opening could be spanned by leaning two stones together, and arches with curved soffits were developed as early as 4000 BC. Arch forms were used by the Greeks, Assyrians and Etruscans, but it was not until Roman times that the arch was fully exploited in bridges and aqueducts. The Romans used semicircular arches, which consisted of stone blocks cut into wedge shapes, building up the complete structure on temporary staging. The development of the arch in medieval and Renaissance times is outlined in the Introduction. The placing of stone voussoirs on temporary staging was also commonly used in the 18th and 19th centuries for arch bridges. Typical examples are Robert Mylne's Blackfriars Bridge over the River Thames (completed in 1769), consisting of nine semi-elliptical arches, and Robert Stephenson's Hutcheson Bridge, Glasgow, completed in 1832.

Construction of Blackfriars Bridge, London, designed by Robert Mylne and completed in 1769. Placing of the voussoirs has been completed for one of the arches.

Sydney Harbour, with Gladesville Bridge in the foreground and Sydney Harbour Bridge in the background.

168

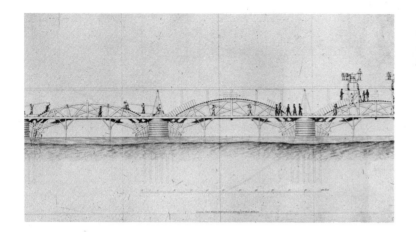

Work in progress on the Hutcheson Bridge, Glasgow (1832), designed by Robert Stephenson. The voussoirs are being placed in position on timber centring.

A large number of concrete arches have been constructed in the 20th century but they are not in general of segmental form. Two notable reinforced-concrete arches cast *in situ* are the Sando Bridge in Sweden of 866-foot span, completed in 1943, and the River Duoro Bridge of 891-foot span built at Oporto in 1963.

The Gladesville Bridge, constructed of hollow concrete voussoirs, has a span of 1000 feet. It crosses the Parramatta River, near the famous Sydney Harbour Bridge, an arch span of 1650 feet constructed in structural steelwork and completed in 1932. The new bridge replaces two earlier swing bridges and carries the North Western Expressway into Sydney.

The bridge was built for the Department of Main Roads, New South Wales. In 1957 tenders were invited for the construction of a steel bridge of cantilever design prepared by the Department. As is common practice, tenderers were allowed to submit alternative designs for large-span bridges. One alternative design submitted by a contracting consortium was for a concrete arch. This design was considered by the Department. During the assessment, the Department was advised by the Professor of Civil Engineering at Sydney University and his staff, and certain design changes were agreed upon. A variation of the alternative design was then proposed, and Eugène Freyssinet and his assistants at the Société Technique pour l'Utilisation de la Precontrainte were asked to report on certain aspects of this design. (This was one of the last projects on which Freyssinet was involved before he died in 1962.) The recom-

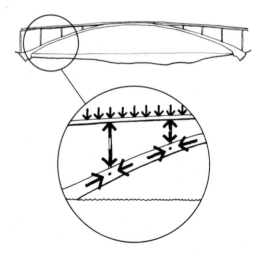

The load from the deck is transmitted to the arch as a series of point loads. To minimise these loads the deck structure was made as light as possible.

170

The dominant member of the structure of Gladesville Bridge is a concrete arch of 1000-foot span. The arch supports the deck structure via a series of prestressed concrete piers. The deck grillage can be clearly seen.

mendations made by Freyssinet in his report were agreed on, and work proceeded in accordance with the revised alternative proposal. The contractors were Stuart Brother and Partner of Sydney, a consortium consisting of Reed and Mallik Ltd. of Salisbury, England, and Stuart Bros. Pty Ltd. of Sydney. The initial and final designs were prepared for the contractors by G. Maunsell & Partners of London and Melbourne, who also designed the formwork.

The general arrangement of the bridge is shown above. The dominant member of the structure is a concrete arch of 1000-foot span. The arch supports the deck structure via a series of prestressed concrete piers. The deck structure continues on each bank of the river and ends at the abutments, which retain the approach embankments.

The load from the deck structure is transmitted to the arch via the prestressed concrete piers as a series of point loads. Since this load was a significant part of the total arch loading it was logical to design the deck structure so that it would not be excessively heavy. The deck consists of a series of 100-foot-span precast, prestressed T-beams. The top flanges and projecting stub diaphragms were connected by *in situ* concrete to form a beam and slab grillage. The deck beams were placed in position by means of a steel launching truss in a similar manner to that adopted for the Medway Bridge. The provision of diaphragms (cross-beams) in the deck is a difficult construction detail but to keep the weight of the deck down, the beams were of fairly slender proportions. So to distribute the effects

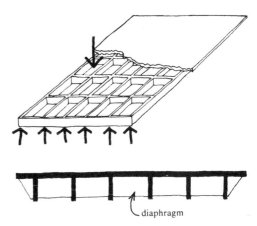

diaphragm

A grillage consisting of longitudinal and cross beams is more efficient than a series of isolated longitudinal beams. The interaction between the two sets of beams means that a concentrated load can be distributed over a number of beams.

171

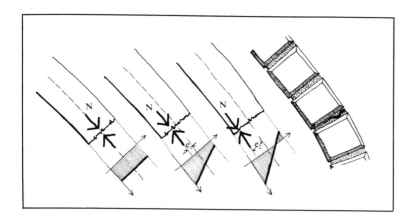

If the thrust acts at the centre of the section the stress distribution is uniform (left).

A small eccentricity (e_1) will induce a non-uniform stress distribution but tension will not occur (centre left).

A larger eccentricity (e_2) will induce tension (centre right).

of a heavy vehicle load acting near the edge of the deck, cross-beams were required. The effect of a heavy vehicle load acting near the edge of a deck is further accentuated if the ratio of the width to the span is large. In a narrow deck the effect is much less pronounced. The deck beams are tapered at the ends to make enough room to cast *in situ* concrete round untensioned reinforcement, so that continuity is provided between adjacent spans. In general, the connection between the deck structure and the piers was by means of a concrete hinge.

It is interesting to compare the deck construction for Gladesville Bridge with that adopted for the viaduct spans for the Medway Bridge. In both bridges, continuity was achieved by means of an *in situ* concrete joint connecting adjacent precast beams. The Medway viaduct beams were deeper and stockier; the extra depth was made necessary by the depth of the adjacent anchor arm. It was therefore possible to avoid the use of intermediate diaphragms (cross-beams). The connection between the Medway viaduct beams and the piers was by means of rollers whereas concrete hinges were used in the Gladesville. This meant that the Gladesville Bridge piers participated in the longitudinal movement of the deck. This was achieved by adopting a very slender column, 2 feet thick, which was prestressed to eliminate tensile stresses. It was necessary to prop the columns during construction to ensure stability. Once again, a comparison of these columns with those adopted for the Medway Bridge viaduct piers is of interest.

The principle of the arch has been outlined, and many engineers from the 17th century onwards put forward methods of analysis.

The Medway Bridge, one of the longest prestressed concrete beam structures in the world.

Rules for the proportioning of masonry arches were available from the 18th century but were intended to apply to small- or medium-span arches designed for light loading. In the past, it was generally believed that the stability of the voussoir arch depended on the avoidance of tension. If we consider a section through a rectangular voussoir arch which is of such a form that the thrust N acts at the centre of the section, the material will be stressed uniformly (above left). If, however, the thrust acts at some distance e_1 from the centre of the section of distribution, stress will no longer be uniform. Further increase in the eccentricity of the thrust to e_2 will produce a tensile stress. For a rectangular section this occurs when the eccentricity e is greater than one sixth of the depth. Engineers considered that since the mortar between the joints was incapable of accepting tensile stress, the joint would be liable to open—in short, it was thought that the stability of the arch was precarious. As has already been shown, if tension occurs at a joint there is still a considerable margin of safety against failure.

The first scientific investigation of the mechanics of the arch was made by Carlo Alberto Pio Castigliano (1847–84). Although of humble background, Castigliano became a brilliant engineer and in 1879 his most famous work was published: *Theory of Equilibrium of Elastic Systems and its Applications*. This book includes a comprehensive theoretical treatment of the arch. It should be noted, however, that treating the arch as an exact inversion of an extensible cable system carrying the same load (see p. 18) is only correct if the voussoirs remain perfectly rigid.

Zones of tension in a concrete arch.

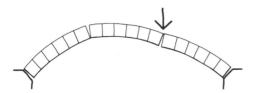

Failure of an arch rib due to the formation of four hinges.

However, the voussoirs will in fact be compressed by the force acting on them, which means the arch will shorten. This will alter the shape of the arch and give rise to bending. However, if the arch rib shortening can be corrected by jacking (described later) the bending can be eliminated.

In his book Castigliano makes a study of several completed structures including the analysis of a masonry bridge over the Doine at Turin, constructed in 1827 by Charles Mosca. The analysis was first carried out on the assumption that the arch was uncracked. The zones of the arch at which tension occurred were removed leaving the arch wholly in compression. The modified arch was analysed in the same manner and the procedure continued until the arch was subjected to compressive stress only. The actual stresses in the masonry determined whether the arch was satisfactory. Castigliano also took into account increased compressibility of the mortar joints compared with the stone.

In 1936 the results of a detailed investigation of the mechanics of the voussoir arch were published by Professor Pippard and Letitia Chitty. The experiments were carried out using a 4-foot span, 1-foot rise segmental arch made up of steel voussoirs. The results of the experiments showed that Castigliano's approach to the analysis of a voussoir arch was sound. Further experiments were made to assess the effect of spread of the abutments. Taking an initially fixed-end arch (that is, where there is no rotational or translational movement of the abutments) a slight spread of the abutments caused the arch to behave as if it were hinged at three points. In general, failure will occur when the structure becomes a mechanism—that is, by the formation of four hinges.

In the Gladesville Bridge, the arch profile was calculated so that under dead load the stress in the voussoirs was almost uniform; the eccentricity of the thrust N was therefore very small. The ratio of dead load to live load was so great that it was not possible to produce tensile stress, and thus failure in the form of a mechanism could not occur. Conversely, the estimated spread of the abutments was so small that the possibility of conversion to a three-hinge arch could be ignored. Another possible form of arch failure is produced by buckling in a way similar to that which occurs in slender columns. The arch is in effect a curved compression member and it is entirely possible that buckling may occur at a load well below that required to crush the material. Another factor that had to be considered in the design of the Gladesville Bridge was the

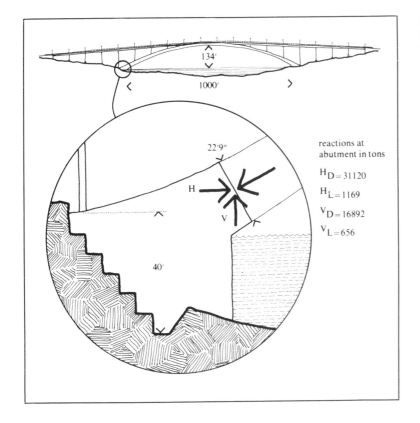

134'

1000'

22'9"

reactions at
abutment in tons

$H_{D} = 31120$

$H_{L} = 1169$

$V_{D} = 16892$

$V_{L} = 656$

H

V

40'

instability during erection both of a single arch rib and of the whole arch, which consisted of four ribs.

A prime consideration in the erection of an arch bridge is the method of supporting the voussoirs during construction. The cost of the staging required to support the arch centering during the placing of the voussoirs represented a substantial proportion of the total construction costs; its design and construction therefore required detailed investigation.

For the Gladesville Bridge a navigation opening of 200 feet by 80 feet was required during construction. The navigation opening was spanned by 220-foot steel trusses with a maximum depth of 28 feet. Steel tubes were used for the staging columns, and the erection was carried out using a 150-ton-capacity floating crane.

Because of congestion at the bridge site, the arch-rib units were cast at nearby Woolwich. The units were cast on end; their dimen-

175

sions varied from 20 feet by 22 feet 7½ inches to 20 feet by 14 feet. Their height varied from 7 feet 7 inches to 10 feet 10½ inches. Each unit was transported to the bridge site by barge. A 50-ton hoist was used to lift the unit up to the arch level. A transporter trolley was then moved under the unit by means of a fixed winch on the upstream side of the hoist. The unit, loaded on the trolley, was then moved into place and lifted off the transporter by means of jacks. It was lowered down the rib and lined up in the correct position. The average rate of placing the units was five to six per day. The 3-inch-wide *in situ* concrete joint between the units was poured as soon as possible after their alignment.

The Department of Main Roads was advised by their consultants to make the top flanges of the units continuous in order to ensure that the four ribs acted as one unit as far as horizontal stability and wind loading were concerned. This was achieved by using *in situ* concrete strips in which were placed prestressing cables. It was also necessary to provide diaphragms under each pier to transmit the lateral load from the deck structure to the four arch ribs. Further diaphragms, one between each pier, were required to prevent the cumulative distortion of the ribs resulting from the moments transmitted from the piers. This effect can be demonstrated by making a thin open-ended cardboard tube and twisting it equally and in opposite directions at each end. This will produce considerable distortion of the tube. But if the ends of the tube are made solid (diaphragms), the twist will induce much less distortion. The distortion will be further reduced by inserting diaphragms at intermediate points. The diaphragm units for the Gladesville Bridge were 2 feet thick.

The final stage in the construction of the arch ribs was the lowering of the centering. If the arch ribs had been cast *in situ*, then the contraction of the arch would have been difficult to predict accurately. However, the *in situ* concrete in the Gladesville arch ribs only amounted to that in the thin joints (3 inches wide). Thus the contraction of the ribs from shrinkage was reduced to a minimum since most of the shrinkage in the voussoirs themselves (which constituted the greater part of the length of the rib) would have taken place during the maturing before erection. The shortening of the voussoirs from the compression to which they were subjected was easier to calculate. The width to span ratio of each arch rib was 1/50 and this did not leave much margin of error for horizontal non-alignment.

Gladesville Bridge, a strikingly modern structure, was built on the same basic principle as the stone bridges of the Romans.

The method adopted by Freyssinet for the Plougastel Bridge was to float out the arch centre (above).

Above left, the temporary structure used to support the Gladesville arch ribs during construction. The navigational opening was spanned by 220-foot steel trusses with a maximum depth of 28 feet.

The Plougastel Bridge over the Elorn (below) was completed in 1930. The arch spans were about 580 feet.

During the jacking, the jacks were placed at the quarter points of the rib. If the arch centering had been removed by simply lowering the staging, there would have been no possibility of correcting for non-alignment. The ribs were therefore sprung from the centering gradually, by jacking about $3\frac{1}{2}$ inches at each quarter point. Three jacks at each end of each vertical were then isolated from the main pumping circuit and were used to adjust the line of action of the thrust and so trim the arch. It was also possible to compensate for any shortening of the arch rib that might have taken place in the way described earlier. The jacking operation completed, it was necessary to transfer the pressure from oil to solid. This was achieved by emptying the jacks of oil one at a time and then pumping in cement paste. In this way, the original jack force was maintained. This method of removing an arch from centering was developed by Freyssinet in the 1920s. He also developed the flat jack used for creating the necessary force.

The Gladesville Bridge is an outstanding example of how a correct assessment of concrete as a structural material has influenced the structural form and the method of construction.

Severn Bridge

The development of the suspension bridge up to the middle of the 19th century has been outlined in the chapter on Brunel's bridge at Clifton. From the middle of the 19th century until the last decade, the construction of long-span suspension bridges was dominated by the Americans. However, in recent years some notable suspension bridges have been constructed in Europe including the Tancarville Bridge in France, the new Forth Bridge, the Tagus Bridge and the Severn Bridge.

America's lead in the design and construction of long-span suspension bridges was largely due to the pioneer work of Charles Ellet and John Augustus Roebling. Charles Ellet was born in 1810 and at the age of twenty went to Paris to study at the Ecole Polytechnique. Thus Ellet obtained first-hand knowledge of the work of French engineers on suspension bridge design and construction. These engineers were Navier, who developed the theory of the behaviour of suspension bridges, Marc Seguin and Gabriel Lamé, who constructed the first wire-rope suspension bridges, and Louis Vicat who invented a method of spinning the cable at the site in 1829.

Ellet put forward many designs for suspension bridges including a bridge crossing the Ohio River at Wheeling, Virginia, which was completed in 1849. This structure had a main span of 1010 feet and the cables consisted of two parallel groups of six containing a total of 6600 wires. The suspenders were inclined slightly to the horizontal and the deck was stiffened by means of longitudinal timber trusses.

The Wheeling Bridge was almost completely destroyed by a storm in 1854. Possible causes of deterioration of the structure were flaws in the wrought-iron wires and inadequate binding of the cables, which meant that the separate strands could not act as one unit and were thus liable to deterioration from friction. John Augustus Roebling was invited to rebuild the bridge and the cable design was improved by compacting the strands and wrapping them in a continuous helix of coated wire. An auxiliary system of suspenders radiating from the stone towers was also adopted.

Roebling was born in Europe and studied at the Royal Polytechnic School in Berlin. He had written a dissertation on the possibilities of the suspension bridge as a student, and in 1837 started his engineering career in America. Roebling constructed the world's first railway suspension bridge over the Niagara River with a span of 821 feet.

The Severn Bridge is the most advanced of its type in the world and is a tribute to the pioneer work of Finley, Telford, Brunel, Ellet, and Roebling.

181

This was a double-deck structure, the upper for trains and the lower for vehicles and pedestrians. The total load was distributed between a system of radiating stays and cables, a common feature of all Roebling's bridges. The construction of a suspension bridge at Pittsburgh in 1860 marked the first use of travelling sheaves to spin the cables. Roebling's greatest bridge work was begun in 1867, a 1595-foot-span structure spanning the East River between Brooklyn and Manhattan. Roebling himself died shortly before construction began but his son Washington Roebling completed the work, which was officially dedicated in 1883.

The Brooklyn Bridge held the record as the world's longest suspension bridge for twenty years, and up to 1929 its span had been exceeded by only 250 feet on completion of the Ambassador Bridge at Detroit. But in the following seven years suspension bridges built before 1930 were to be dwarfed by the construction of the George Washington Bridge spanning the Hudson at New York City (main span 3500 feet) and the Golden Gate Bridge at San Francisco (main span 4200 feet).

The collapse of the Tacoma Narrows Bridge at Puget Sound in 1940 led to the immediate investigation of the susceptibility of suspended structures to wind-excited oscillations. The work was started at the University of Washington with the aim of determining the mechanism of the wind action, which badly distorted suspension bridges, and of assessing the stability of the proposed design for the new Tacoma Narrows Bridge, which was completed in 1950. There was also some concern about the movements of the Golden Gate Bridge, which led to further bracing being added to the stiffening girder.

Before the Second World War Messrs Mott, Hay & Anderson, consulting engineers, prepared a report on the need for a road bridge crossing the River Severn and linking the industrial centres of South Wales and Southern England. Proposals for a bridge were shelved during the war but in 1945 the Ministry of Transport assumed responsibility for a project to bridge the Severn. In 1946 Messrs Mott, Hay & Anderson and Freeman Fox and Partners were engaged as joint consulting engineers. The Ministry of Transport built a wind tunnel at Thurleigh, near Bedford, in 1947 to carry out aerodynamic tests on a model of the proposed structure. The data obtained was also to be used for the proposed Forth Road Bridge, which was to be of similar span. At this time it was thought that the same deck design might be adopted for both structures. The model

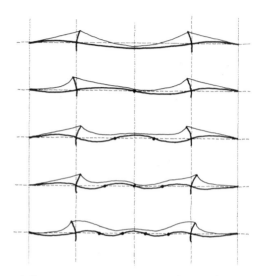

A diagrammatic representation of vertical movement of a suspension bridge deck. This type of movement is not in general as critical as twisting about a span-wise axis.

Brooklyn Bridge, span 1595 feet between the towers, designed by J. A. Roebling. Note the use of vertical and radiating suspenders, a common feature of Roebling's bridge designs.

tests and full-scale experience showed that the oscillations in suspension bridges produced by steady winds take the following forms: first, vertical motion, in which the deck structure moves up and down. This type of movement is rarely dangerous but would cause some concern to motorists crossing the bridge. Second, torsional twisting motion in which the deck twists about a spanwise axis, the cables displacing by an equal amount but in opposite directions. It was this type of motion that caused the collapse of the Tacoma Narrows Bridge. It was found that bridges stiffened by plate girders (for example, the first Tacoma Bridge) were susceptible to both types of motion, whereas bridges stiffened by trussed girders (for example, the Forth Bridge and the Verrazano Narrows Bridge) were largely immune to vertical motion and that instability in torsional motion could be corrected by a suitable design of the suspended structure.

The essential features of the stiffening girder should be as follows: the ratio of the depth of the main truss to the width should be high; the carriageways should be separated by openings or gratings; the secondary members should also be of the truss type and not solid; the parapets should not be solid; the footpaths, cycle tracks and so on should be placed outside the main trusses; bracing in the plane of the carriageways should be placed near deck level.

After work on the Forth Bridge was started in 1958, the consulting engineers decided to change the design of the Severn Bridge deck, adopting a shallow steel-plated box structure. The final profile, the result of the extensive model tests, derived its stability from the aerodynamic shape of an extremely shallow structure only 10 feet deep. The depth of the Forth Bridge stiffening trusses was 27 feet 6 inches. The Severn Bridge uses an inclined suspender system, thus reducing the likelihood of vertical motions, which would be liable to occur in a shallow deck construction.

The Severn Bridge is sited about eight miles upstream from Avonmouth between the Aust Cliff and Beachley Peninsula. At this point the River Severn is about one mile wide at high water. The bridge forms part of the M4 Motorway from London to South Wales. The Firth of Forth site, which is always navigable, differs considerably. The tidal range of the Severn estuary is 46 feet, one of the largest in the world, and at low tide reveals sandbanks and rocks. These conditions posed construction problems, which will be described later. Work on the substructure (piers supporting the towers and anchorage blocks) was started in May 1961.

184 The construction of the east (Aust) pier was a slow process as

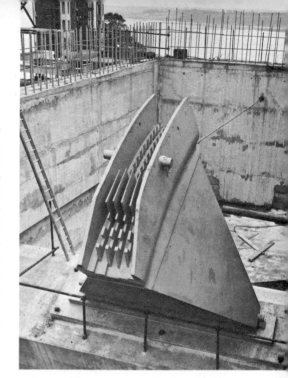

The Beachley downstream splay saddle in position. The groove in the top of the saddle carries one end of the main cable at the point where the individual strands splay out to be secured at the bottom of the anchorage block.

The east (Aust) pier, left, and the west (Beachley) pier. The latter is supported on two concrete cylinders which extend about 33 feet down from the river bed.

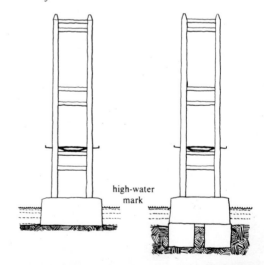

high-water mark

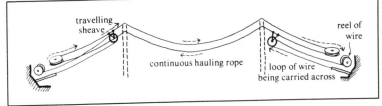

travelling
sheave

reel of
wire

continuous hauling rope

loop of wire
being carried across

Work in progress on the Beachley tower, showing two of the three horizontal cross members which brace the box-section vertical members.

The catwalk and spinning wheel leaving the Beachley tower on the downstream side.

A diagrammatic representation of the cable-spinning process.

the men only worked for a short time each day at low tide. The pier was founded at a level 18 feet below Ordnance Datum on a limestone outcrop in the river. It was stressed down into the rock by high-tensile alloy bars and was constructed by placing precast concrete blocks to form an outer skin and then filling them with concrete poured *in situ*. The pier reaches a height of 63 feet above its foundation level and the plan dimensions are 132 feet by 63 feet. It has pointed cutwater ends (see p. 30). The west (Beachley) pier is similar to the Aust pier but is supported on two 60-foot-diameter concrete cylinders extending about 33 feet down from the river bed to found on red marlstone at 50 feet below Ordnance Datum.

The anchorage blocks appear externally as large rectangular blocks of concrete 155 feet long by 100 feet wide and reaching a height of 132 feet above Ordnance Datum. The weight of the anchorage had to be great enough to resist the pull from the two cables, which was 19,200 tons under dead load. The blocks have hollow 185 interiors and there is an access shaft for inspecting the cables. Inside the front part of the anchorage block 'splay saddles' were constructed. Their purpose was to spread the cables into separate strands, each of which has a separate anchorage in the block.

Erection of the superstructure (towers, cables and suspended structure) was started in March 1963. The first stage of the work was the erection of the steel towers. Previous experience on the Forth Bridge had shown that the towers would be susceptible to bending oscillations during construction, which would hamper progress of the work and possibly cause structural damage.

Model tests indicated that the oscillations would be induced by winds blowing in line with the two legs of the tower at speeds of 20 and 80 mph. To prevent oscillation building up, the towers were restrained by wire cables, which limited the movements to 2 inches.

To erect the towers a 'climbing structure' was designed, which was raised in stages as the work progressed. The towers consist of two slightly tapering rectangular box sections, 400 feet high and connected by three deep horizontal cross-members. The boxes are braced internally about every 55 feet by horizontal members and also at the points of attachment of cross-members. The tower design is remarkable for the low tonnage of the steel employed, about one half of that used for the Forth Bridge towers.

The construction of the cables was by the 'spinning process', which has been used in America for about a century. The process consists of building up the cable from a large number of single wires, which are laid parallel to one another and compacted into a large bundle of circular cross-section. This produces a cable of minimum cross-section for the load to be carried. (An alternative procedure is to preform the wire into separate strands. This was adopted for the Tancarville Bridge with a main span of 2000 feet. Each of the main cables consist of 60 separate $2\frac{3}{4}$-inch-diameter strands, which were erected by hauling them across the river one at a time. In the future this method may well be applied to larger spans.)

The first stage of the cable-spinning process was to erect a catwalk a few feet below the intended line of the cable. The main cable wires (0·196 inch in diameter) were reeled on to 6-foot-diameter steel drums. The end of a wire was pulled off the wire reel and looped round a spinning wheel, attached to the hauling rope, which was pulled over the towers to the opposite anchorage. There, the wire loop was slipped off the spinning wheel and placed on a strand shoe. 186 At this stage another wire loop was placed on the spinning wheel,

A 60-foot length of deck section being towed into position (top).

The cable bands are clamped to the cables by high-tensile steel threaded rods. The two suspenders are connected to the bottom half of the cable band.

Erection of the deck started from the centre of the main span and worked back towards the piers.

which returned with it to the starting anchorage. The process was continued until the spinning of a strand was completed. Each cable was built up in the form of 220 loops. At the anchorages, each strand was looped round the semicircular mild-steel strand shoes, which were held back by two 4¼-inch-diameter high-tensile steel rods to further steelwork embedded in the concrete. Outside the anchorages the cables were compacted into a circular shape and wrapped by machine over wet red-lead paste; the outside of the wrapping wire was painted.

The sag of the cables in relation to the span (dip) affects the height of the towers, the pull on the anchorages and the stiffness of the structure as a whole. For minimum cable stress the dip should be large, but taking into account all the factors involved, a ratio of 1:12 was adopted.

The inclined suspenders supporting the deck were just over 2 inches in diameter and consisted of single strands containing 178 galvanised steel wires. The strands were socketed at each end, the upper ends of two suspenders being connected to cable bands, which were clamped to the main cables at 60-foot centres. The cable bands consisted of steel castings, split on a horizontal diameter and

187

The Verrazano Narrows Bridge, New York, the world's longest suspension span of 4260 feet, which carries 12 lanes of traffic.

clamped to the cable by high-tensile steel-threaded rods.

The deck structure consisted of a hollow steel box 10 feet deep by 75 feet wide, which was prefabricated in 60-foot sections. The sections were launched into the River Wye and towed to the site for hoisting into position. Again, extensive model tests were required to assess the floating and towing capabilities of the deck sections.

The operation of positioning the sections under the bridge and lifting them into place depended on the tides. The rate of progress was considerably restricted by the unusual tidal conditions, and in a less hazardous stretch of water the work would have proceeded much faster. Erection started from the centre of the main span working back towards the piers. When the welding of the suspended deck units was complete, the remaining work consisted of surfacing the roadways, erecting crash barriers, fixing the lighting and final painting. The Severn Bridge was officially opened on September 8, 1966. Faced with a very difficult site, the consultants and contractors produced a structure that is technically the most advanced of its type in the world, and is a fitting tribute to pioneer work of engineers such as Finley, Telford, Brunel, Ellet and Roebling, whose foresight and courage is a source of inspiration to all bridge engineers.

Further reading list

Condit, C. W., *American Building Art - the 19th Century;* New York, 1960

De Maré, Eric, *The Bridges of Britain;* London, 1954

Mock, Elizabeth, *The Architecture of Bridges;* New York, 1949

Pannell, J. P. M., *An Illustrated History of Civil Engineering;* London, 1964

Shirley-Smith, H., *The World's Great Bridges;* rev. ed., London, 1964

Steinman, D. B., and Watson, S. R., *Bridges and their Builders;* New York, 1957

Thul, H., *Cable-Stayed Bridges in Germany;* B.C.S.A. paper, London, 1966

Timoshenko, Stefan, *The History of Strength of Materials;* New York, 1953

Glossary

abutment: an end support to a bridge which carries load from deck to ground.

arch: a curved structural member acting principally in compression.

beam: a straight structural member which acts principally in bending and shear; for vertical loading the beam supports resist vertical forces only.

bearing: a device for transmitting the load from a beam to its support.

bending moment: the bending effect at any section of a structure due to the applied loads; it can be expressed as force times distance.

box beam: a hollow beam or girder, designed to give strength without great weight.

brace: a stiffening member in a structure, usually diagonal and normally designed to resist wind.

caisson: a circular or rectangular structure which prevents water flowing into an excavation during construction of bridge foundations.

cantilever: a beam free at one end and built in or counterbalanced at the other.

cast iron: an alloy of iron and carbon containing about 2 per cent carbon; notoriously unreliable in tension.

cofferdam: a temporary dam, usually of timber or steel, giving access to ground that is normally submerged.

compression: a force action which tends to shorten a member (the opposite of tension).

creep: the gradual deformation of a material, particularly concrete, under load.

formwork: the entire system required to support and form concrete members.

force: that which tends to produce movement in a body; e.g. the weight of a body is a 'force' which tends to move it downwards.

parabola: the curve assumed by a cable of negligible weight when subjected to uniformly distributed loading.

pier: an intermediate support for a bridge, normally in the form of a short wall, which divides the deck into two or more spans.

pile: a structural element which is driven or cast into the ground to transmit the loads from the structure to firmer soil or rock below ground level.

pile cap: the concrete cast around the top of a group of piles which ensures that they act together in transmitting the load from the structure.

precast concrete: a concrete member cast before being placed in position.

prestressed concrete: concrete in which a state of compression has been induced prior to the application of loads.

reaction: the resistance of a support, such as a pier or abutment to the loading from the bridge deck.

reinforced concrete: concrete in which steel or other material has been incorporated to resist tensile stresses.

shear: a force action which produces a racking effect.

shrinkage: shortening of concrete during the hardening process.

strain: the ratio of change in length to original length of a member under load.

stress: the intensity of load per unit area on a member.

tension: a force action which tends to extend a member (the opposite of compression).

torsion: a force action which tends to twist a member.

truss: an assembly of members normally designed to transmit loads by means of direct tension and compression with no bending.

wrought iron: iron with a very low carbon content; less brittle than cast iron.

Index

Figures in italics refer to illustrations

190

Acknowledgements

Key to picture positions: (T) top, (C) centre, (B) bottom, (L) left, (R) right. Numbers refer to the pages on which pictures appear.

Aargauische Denkmalpflege 89(T); Acier-Stahl-Steel 158; Aerofilms 44; R. S. Allen Coll. New York 91; Associated Press 18; Australian News and Information Bureau 171; Barnaby's Picture Library, 41, 45, 101; Julius Berger AG, Wiesbaden, 149, 150, 151, 152, 153, 154, 155(T & B); Bibliothèque Nationale 34(B); Jean Bottin 24, 98; Britain-China Friendship Assoc. 27; British Constructional Steelwork Assoc. 159, 161; British Museum 28, 168; British Railways 90; Bristol City Art Gallery 68(T & B) 70; Bristol University Library 6(B); Camera Press 33, 188; J. Allan Cash 25(B), 29, 47, 75, 123; Cement and Concrete Assoc. 8, 42 (T & C), 48, 50, 51, 81(T & B), 127, 145(T), 178(B); Deutsches Museum, München 89 (B); Henry L. Doten, Augusta, Maine 94(B); Dr. Llewellyn N. Edwards 94(T); design study by Dr. A. R. Flint 9; Freeman Fox & Partners 173; L & M Gayton 55 (T); George H. Hall, Cardiff 180; The Hamlyn Group Picture Library 57, 58, 59, 60, 61, 62, 63, 92(T & B), 93, 97, 112(B), 118, 119, 120, 129, 133; Bétons Armés Hennebique 125; prof. arch. Italo Insolera, Roma 102, 103, 105, 106(T & B), 109 (T & B); A. F. Kersting 64, 77, 104; Keystone Press Agency Ltd. 111; Ralph Kleinhempel, Hamburg 164; The Mansell Collection 6(T), 26, 112(T), 115, 118; Eric de Maré 22, 23, 30, 31(T), 37, 40, 55(C), 71, 87, 117; Foto Marburg 163(T, L & R); Maryland Dept. of Economic Development 147; G. Maunsell & Partners 169, 178(T); Ministerio de Obras Publicas, Caracas Frontispiece; National Monuments Record 53, 54, 55(B); Popperfoto 25(T), 31(B), 32(B), 183; Radio Times Hulton Picture Library 32(T), 69, 79, 85, 170; Rapho 35; Réalités 107; Rheinstahl Union AG, Dortmund 157, 165, 166(T), 166(B); H. Roger-Viollet 39, 42(B), 43; St. Louis-San Francisco Railway Co. 100(T); Scottish Tourist Board 100(B); Science Museum, London 34 (T), 38, 67, 82, 183; Société Technique pour l'Utilisation de la Précontrainte 137, 139, 140(T & B), 143 (T & B), 179; State Highway Dept. Connecticut 131 (T & B); Swiss National Tourist Office 126; Ministerialrat Prof. H. Thul, Bonn, 166(C), 167; William Tribe Ltd. 184, 185, 186, 187; U.S. Information Services 46; Reece Winston, Bristol 72(T & B), 73, 74; Z.F.A. 177.

The author and publishers wish to thank in addition Mr. Richard Lyons who produced the diagrams and explanatory drawings for the text.